WILEY SERIES IN COMMUNICATION AND DISTRIBUTED SYSTEMS

VSAT NETWORKS

G. Maral

Ecole Nationale Supérieure des Télécommunications (Telecom Paris)
site of Toulouse, France

JOHN WILEY & SONS

Chichester · New York · Brisbane · Toronto · Singapore

Other Wiley Editorial Offices

John Wiley & Sons, Inc., 605 Third Avenue,
New York, NY 10158-0012, USA

Jacaranda Wiley Ltd, G.P.O. Box 859, Brisbane,
Queensland 4001, Australia

John Wiley & Sons (Canada) Ltd, 22 Worcester Road,
Rexdale, Ontario M9W 1L1, Canada

John Wiley & Sons (SEA) Pte Ltd, 37 Jalan Pemimpin #05-04,
Block B, Union Industrial Building, Singapore 2057

Library of Congress Cataloging-in-Publication Data

Maral, Gérard.
    VSAT networks / G. Maral.
        p.   cm.—(Wiley series in communication and distributed
    systems)
    Includes bibliographical references and index.
    ISBN   0 471 95302 4 : $39.95 (U.S.)
    1. VSATs (Telecommunication)   I. Title.   II. Series.
TK5104.2.V74M37   1995   94-37789
384.5'1—dc20   CIP

British Library Cataloguing in Publication Data

A catalogue record for this book is available from the British Library

ISBN   0 471 95302 4

Typeset in $10\frac{1}{2}/12\frac{1}{2}$ pt Palatino by Thomson Press (India) Ltd., New Delhi
Printed and bound in Great Britain by Bookcraft (Bath) Ltd

CONTENTS

PREFACE

Satellites for communication services have evolved quite significantly in size and power since the launch of the first commercial satellites in 1965. This has permitted a consequent reduction in the size of earth stations, and hence their cost, with a consequent increase in number. Small stations, with antennas in the order of 1.2–1.8 m, have become very popular under the acronym VSAT, which stands for 'Very Small Aperture Terminals'. Such stations can easily be installed at the customer's premises and, considering the inherent capability of a satellite to collect and broadcast signals over large areas, are being widely used to support a large range of services. Examples are broadcast and distribution services for data, image, audio and video, collection and monitoring for data, image and video, two-way interactive services for computer transactions, data base inquiry, and voice communications.

The trend towards deregulation, which started in the United States, and is presently progressing in other regions of the world, has triggered the success of VSAT networks for corporate applications. This illustrates that technology is not the only key to success. Indeed, VSAT networks have been installed and operated only in those regions of the world where demand existed for the kind of services that VSAT technology could support in a cost effective way, and also where the regulatory framework was supportive.

The success of the book *Satellite Communications Systems* by myself and Michel Bousquet (2nd edition, Wiley, 1993) was an encouragement to develop in a new book those aspects which are specific to VSAT networks. I also was pushed by my students, either the younger ones from TELECOM Paris, where this material has been elaborated during dedicated workshops, or the elder and also more mature ones, who attend my continuing education courses on various aspects of satellite communications, hosted by CEI-Europe.

The present book on VSAT networks aims at introducing the reader to the important issues of services, economics and regulatory aspects. It is also intended to give detailed technical insight on networking and radiofrequency link aspects, therefore addressing the specific features of VSAT networks at the three lower layers of the OSI Reference Layer Model for data communications.

From my experience in teaching, I felt I should proceed from the general to the particular. Therefore, Chapter 1 can be considered as an introduction to the subject, with rather descriptive contents on VSAT network configurations, services, operational and regulatory aspects.

The more intrigued reader can then explore the subsequent chapters. Chapter 2 deals with those aspects of satellite orbit and technology which influence the operation and performance of VSAT networks. Chapter 3 gives more details on the regulatory and operational aspects which are important to the customer. Licensing is discussed, installation problems are presented, and a list of potential concerns to the customer is explored. Hopefully, this chapter will not be perceived as discouraging, but on the contrary as a friendly guide for avoiding misfortunes, and getting the best from a VSAT network.

The next two chapters are for technique oriented readers. Actually, I thought this would be a piece of cake for my students, and a reference text for network design engineers. Chapter 4 deals with networking. It introduces traffic characterisation, and discusses network and link layers protocols of the OSI Reference Layer Model, as used in VSAT networks. It also presents simple analysis tools for the dimensioning of VSAT networks from traffic demand and user specifications in terms of blocking probability and response time. Chapter 5 covers the physical layer, providing the basic radio frequency link analysis, and presenting the parameters that condition link quality and availability. An important aspect discussed here is interference, as a result of the small size of the VSAT antenna, and its related large beamwidth.

Appendices are provided for the benefit of those readers who may lack some background and have no time or opportunity to refer to other sources. I also felt the reader would appreciate finding here a description of some of the existing VSAT products. Although this information is subject to rapid change, and does not cover all available products, it is interesting for comparing today's offers. Actually, the offer from one vendor to another on technical grounds is quite similar.

Many people have helped me during the preparation of this book. First of these is Helge Grünbaum, from the National Post and Telecom Agency of Sweden, who reviewed the manuscript, proposing significant improvements. I would also like to thank those people who provided me with useful documentation and advice: Jonathan Collins and Richard Barrett from ATT Tridom, Rami Kaoua from GTE Spacenet, Helio Martins from HNS France, Yasuhiko Ito from KDD R&D Labs, Christophe von Rakowski and Gerard Elineau from Dornier, Ian Westall from Ferranti International, Nissan Leviathan from Gilat Europe, J.D. Rogers from Multipoint Communications, Sven-Erik Hennum-Johnsen from Normac, Didier Chaminade from France Cables and Radio, Jean-Luc Garneau from Polycom, Bernard Guinel, Daniel Lagrange and Patrice Paquot from Eutelsat, and Bernard Lancrenon and Eric Alberty from Matra Marconi Space. I would also like to mention one of my students, Jean-Claude Pommares. Discussions with people of TELECOM Paris were also of great help, namely Dominique Seret from the Networks Department, and Tarif Zein, Pascal Lalanne and Alain Pirovano, from the Toulouse site. Some of the material in the book has been inspired by the fruitful work performed within the COST226 project, chaired with talent by my colleague Professor Otto Koudelka from the University of Graz.

Finally, many thanks to all the students I have taught, at TELECOM Paris, the University of Surrey, CEI-Europe, and other places, who, by raising questions, asking for details and bringing in their comments, have helped me to organise the material presented here.

Writing a book is a long effort, and must be carried out along with day-to-day work. The only place I could find long periods of quietness for thinking and writing has been at home. I would like to thank my wife Elena and my son Roman for accepting me to stay up late at night typing on my word processor, and spending most of our vacation period working on the manuscript rather than in their company. I appreciate their support and understanding.

Gérard Maral

ACRONYMS AND ABBREVIATIONS

ABCS — Advanced Business Communications via Satellite

ACI — Adjacent Channel Interference

ACK — ACKnowledgement

AMP — AMPlifier

ARQ — Automatic repeat ReQuest

ARQ-GB(N) — Automatic repeat ReQuest—Go Back N

ARQ-SR — Automatic repeat ReQuest—Selective Repeat

ARQ-SW — Automatic repeat ReQuest—Stop and Wait

ASYNC — ASYNChronous data transfer

BEP — Bit Error Probability

BER — Bit Error Rate

BITE — Built-In Test Equipment

BPF — Band Pass Filter

BPSK — Binary Phase Shift Keying

BSC — Binary Synchronous Communications (bisync)

CCI — Co-Channel Interference

CCIR — Comité Consultatif International des Radiocommunications (International Radio Consultative Committee)

CCITT — Comité Consultatif International du Télégraphe et du Téléphone (International Telegraph and Telephone Consultative Committee)

CCU — Cluster Control Unit

CDMA — Code Division Multiple Access

CFDMA — Combined Free/Demand assignment Multiple Access

CFRA — Combined Fixed/Reservation Assignment

COST — European COoperation in the field of Scientific and Technical research

DA — Demand Assignment

DAMA — Demand Assignment Multiple Access

dB — deciBel

D/C — Down-Converter

DCE — Data Circuit Terminating Equipment

DEMOD — DEMODulator

DTE — Data Terminating Equipment

ECU — European Currency Unit

EIA — Electronic Industries Association

EIRP — Effective Isotropic Radiated Power

$EIRP_{ES}$ — Effective Isotropic Radiated Power of earth station (ES)

$EIRP_{SL}$ — Effective Isotropic Radiated Power of satellite (SL)

ES — Earth Station

ETR — ETSI Technical Report

ETS — European Telecommunications Standard, created within ETSI

ETSI	European Telecommunications Standards Institute
EUTELSAT	EUropean TELecommunications SATellite Organisation
FA	Fixed Assignment
FCC	Federal Communications Commission, in the USA
FDM	Frequency Division Multiplex
FDMA	Frequency Division Multiple Access
FEC	Forward Error Correction
FET	Field Effect Transistor
FIFO	First In First Out
FODA	FIFO Ordered Demand Assignment
FSK	Frequency Shift Keying
GBN	Go Back N
HDLC	High level Data Link Control
HEMT	High Electron Mobility Transistor
HPA	High Power Amplifier
IAT	InterArrival Time
IBO	Input Back-Off
IDU	InDoor Unit
IF	Intermediate Frequency
IM	InterModulation
IMUX	Input Multiplexer
IP	Internet Protocol
IPE	Initial Pointing Error
ISDN	Integrated Services Digital Network
ISO	International Organisation for Standardisation
ITU	International Telecommunication Union
LAN	Local Area Network
LAP	Link Access Protocol
LNA	Low Noise Amplifier
LO	Local Oscillator
MAC	Medium Access Control, also Multiplexed Analog Components

MCPC	Multiple Channels Per Carrier
MIFR	Master International Frequency Register
MOD	MODulator
MTBF	Mean Time Between Failures
MX	MiXer
MUX	MUltipleXer
NACK	Negative ACKnowledgement
NMS	Network Management System
OBO	Output Back-Off
ODU	OutDoor Unit
OMUX	Output MUltipleXer
OSI	Open System Interconnection
PABX	Private Automatic Branch eXchange
PAD	Packet Assembler/Disassembler
PBX	Private (automatic) Branch eXchange
PC	Personal Computer
PDF	Probability Density Function
PDU	Protocol Data Unit
POL	POLarization
PSD	Power Spectral Density
PSK	Phase Shift Keying
QPSK	Quaternary Phase Shift Keying
RCVO	ReCeiVe Only
Rec	Recommendation
Rep	Report
RF	Radio Frequency
RTT	Round Trip Time
RX	Receiver
S-ALOHA	Slotted ALOHA protocol
SCADA	Supervisory Control And Data Acquisition
SCPC	Single Channel Per Carrier
SDLC	Synchronous Data Link Control
SKW	Satellite-Keeping Window
SL	SateLlite

SNA	Systems Network Architecture (IBM)	TTC	Telemetry, Tracking and Command
SNG	Satellite News Gathering	TV	TeleVision
SR	Selective Repeat	TWT	Travelling Wave Tube
SSPA	Solid State Power Amplifier	TX	Transmitter
SW	Stop and Wait		
		VSAT	Very Small Aperture Terminal
TCP	Transmission Control Protocol		
		XPD	Cross Polarisation Discrimination
TDM	Time Division Multiplex		
TDMA	Time Division Multiple Access	XPI	Cross Polarisation Isolation

NOTATION

A attenuation (larger than one in absolute value, therefore positive value in dB), also length of acknowledgement frame (bits)

A_{RAIN} attenuation due to rain

Az azimuth angle (degree)

a semi-major axis (m)

B bandwidth (Hz)

B_i interfering carrier bandwidth (Hz)

B_{inb} inbound carrier bandwidth (Hz)

B_N receiver equivalent noise bandwidth (Hz)

B_{outb} outbound carrier bandwidth (Hz)

B_{Xpond} transponder bandwidth (Hz)

BU burstiness

c speed of light: $c = 3 \times 10^8$ m/s also distance from centre of ellipse to its focus (m)

C carrier power (W)

C_D carrier power at earth station receiver input (W)

C_U carrier power at satellite transponder input (W)

C_x received carrier power on X-polarisation (W)

C_y received carrier power on Y-polarisation (W)

C/N carrier to noise power ratio

$(C/N)_D$ downlink carrier to noise power ratio

$(C/N)_{Dsat}$ same as above, at saturation

$(C/N)_{IM}$ carrier to intermodulation noise power ratio (Hz)

$(C/N)_U$ uplink carrier power to noise power ratio

$(C/N)_{Usat}$ same as above, at saturation

$(C/N)_T$ overall link (from station to station) carrier to total noise power ratio

C/N_i carrier to interference power ratio

$(C/N_i)_D$ downlink carrier to interference power ratio

$(C/N_i)_U$ uplink carrier to interference power ratio

$(C/N_i)_T$ overall link (from station to station) carrier to interference power ratio

C/N_0 carrier power to noise power spectral density ratio (Hz)

$(C/N_0)_D$ downlink carrier power to noise power spectral density ratio (Hz)

$(C/N_0)_{Dsat}$ same as above, at saturation (Hz)

$(C/N_0)_{IM}$ carrier power to intermodulation noise power spectral density ratio (Hz)

$(C/N_0)_U$ uplink carrier power to noise power spectral density ratio (Hz)

$(C/N_0)_{Usat}$ same as above, at saturation (Hz)

$(C/N_0)_T$	overall link (from station to station) carrier power to total noise power spectral density ratio (W/Hz)
C/N_{0i}	carrier power to interference noise power spectral density ratio (Hz)
$(C/N_{0i})_D$	downlink carrier power to interference noise power spectral density ratio (Hz)
$(C/N_{0i})_U$	uplink carrier power to interference noise power spectral density ratio (Hz)
$(C/N_{0i})_T$	overall link (from station to station) carrier power to total interference noise power spectral density ratio (W/Hz)
D	antenna diameter (m), also number of data bits per frame to be conveyed from source to destination
dBx	value in dB relative to x
E	elevation angle (degree), also energy per bit (J)
E_b	energy per information bit (J)
E_c	energy per channel bit (J)
e	eccentricity
EIRP	effective isotropic radiated power of transmitting equipment (W)
$EIRP_{ES}$	EIRP of earth station (W)
$EIRP_{ESmax}$	maximum value of $EIRP_{ES}$ (W)
$EIRP_{ESsat}$	value of $EIRP_{ES}$, at transponder saturation (W)
$EIRP_{ESi}$	EIRP of interfering earth station (W)
$EIRP_{ESi,max}$	maximum value of earth station EIRP allocated to interfering carrier (W)
$EIRP_{ESw}$	EIRP of wanted earth station (W)
$EIRP_{SL}$	EIRP of satellite transponder (W)
$EIRP_{SLsat}$	EIRP of satellite transponder at saturation (W)

$EIRP_{SL1sat}$	EIRP of satellite transponder in beam 1 at saturation (W)
$EIRP_{SL2sat}$	EIRP of satellite transponder in beam 2 at saturation (W)
$EIRP_{SLi,max}$	maximum value of interfering satellite EIRP allocated to interfering carrier (W)
$EIRP_{SLw,max}$	maximum value of wanted satellite EIRP for wanted carrier (W)
$EIRP_{SLww}$	wanted satellite EIRP for wanted carrier in direction of wanted station (W)
$EIRP_{SLiw}$	interfering satellite EIRP for interfering carrier in direction of wanted station (W)
$EIRP_{SL1ww}$	EIRP of satellite transponder in beam 1 for wanted carrier in direction of wanted station (W)
$EIRP_{SL2iw}$	EIRP of satellite transponder in beam 2 for interfering carrier in direction of wanted station (W)
$EIRP_{SL1wsat}$	EIRP of satellite transponder in beam 1 in direction of wanted station at saturation (W)
$EIRP_{SL2wsat}$	EIRP of satellite transponder in beam 2 in direction of wanted station at saturation (W)
f	frequency (Hz): $f = c/\lambda$
f_D	downlink frequency (Hz)
f_{IM}	frequency of an intermodulation product (Hz)
f_{LO}	local oscillator frequency (Hz)
f_U	uplink frequency (Hz)
G	power gain (larger than one in absolute value, therefore positive value in dB), also normalised offered traffic, also gravitational constant: $G = 6.672 \times 10^{-11} \, \text{m}^3/\text{kg} \, \text{s}^2$
G_{cod}	coding gain (dB)
G_D	power gain from transponder output to earth station receiver input

G_{IF}	intermediate frequency amplifier power gain
G_{LNA}	low noise amplifier power gain
G_{max}	maximum gain
G_{MX}	mixer power gain
G_R	antenna receive gain in direction of transmitting equipment
G_{Rmax}	antenna receive gain at boresight
G_{RX}	receiving equipment composite receive gain: $G_{RX} = G_{Rmax}/L_R L_{pol} L_{FRX}$
G_{RXmax}	maximum value of G_{RX}
G_{RXi}	receiving equipment composite receive gain for interfering carrier
G_{RXw}	receiving equipment composite receive gain for wanted carrier
G_T	antenna transmit gain in direction of receiving equipment
G_{Tmax}	antenna transmit gain at boresight
$G_{Ti,max}$	antenna transmit gain at boresight for interfering carrier
G_{T1w}	satellite beam 1 transmit antenna gain in direction of wanted station
G_{T2w}	satellite beam 2 transmit antenna gain in direction of wanted station
G_{TE}	power gain from satellite transponder input to earth station receiver input
G_{Xpond}	transponder power gain
G_1	gain of an ideal antenna with area equal to 1 m²: $G_1 = 4\pi/\lambda^2$
G/T	figure of merit of receiving equipment (K^{-1})
$(G/T)_{ES}$	figure of merit of earth station receiving equipment (K^{-1})
$(G/T)_{ESmax}$	maximum value of $(G/T)_{ES}$
$(G/T)_{SL}$	figure of merit of satellite receiving equipment (K^{-1})
H	total number of bits in the frame header (and trailer if any)

i	orbit inclination
I_x	received cross polar interference on X-polarisation (W)
IBO	input back-off
IBO_{inb}	input back-off for inbound carrier
IBO_{outb}	input back-off for outbound carrier
IBO_1	input back-off per carrier with multicarrier operation mode
IBO_t	total input back-off with multicarrier operation mode
J_x	cross polar interference on X-polarisation generated by receive antenna (W)
k	Boltzmann constant: $k = 1.38 \times 10^{-23}$ J/K; k (dBJ/K) $= 10 \log k = -228.6$ dBJ/K
l	Earth station latitude with respect to the satellite latitude (degrees)
L	loss (larger than one in absolute value, therefore positive value in dB), also earth station relative longitude with respect to a geostationary satellite (degrees), also length of a frame (bits), also length of a message (bits)
L_a	Earth station relative longitude with respect to the adjacent satellite (degrees)
L_w	Earth station relative longitude with respect to the wanted satellite (degrees)
L_D	downlink path loss
L_{FRX}	feeder loss from antenna to receiver input
L_{FTX}	feeder loss from transmitter output to antenna
L_{pol}	antenna gain loss as a result of antenna polarisation mismatch

L_R	off-axis receive gain loss		P_x	transmitted carrier power on X-polarisation (W)
L_{Rmax}	maximum value of L_R			
L_U	uplink path loss		P_y	transmitted carrier power on Y-polarisation (W)
L_{Ui}	Uplink path loss for interfering carrier		PSD	power spectral density (W/Hz)
			PSD_i	interfering carrier power spectral density (W/Hz)
L_{Uw}	Uplink path loss for wanted carrier			
			PSD_w	wanted carrier power spectral density (W/Hz)
M_e	mass of the Earth: $M_e = 5.974 \times 10^{24}$ kg			
N	noise power (W)		Q_x	cross polar interference on X-polarisation generated by transmit antenna (W)
N_i	interference power (W)			
N_{IM}	intermodulation noise power (W)			
N_{0D}	downlink thermal noise power spectral density (W/Hz)		r	distance from centre of earth to satellite
N_{0U}	uplink thermal noise power spectral density (W/Hz)		R	range, also bit rate
			R_a	slant range from earth station to adjacent satellite
N_{0iD}	downlink interference power spectral density (W/Hz)		R_b	information bit rate (b/s)
N_{0IM}	intermodulation noise power spectral density (W/Hz)		R_{binb}	information bit rate on inbound carrier (b/s)
N_{0iU}	uplink interference power spectral density (W/Hz)		R_{boutb}	information bit rate on outbound carrier (b/s)
N_{0T}	total noise power spectral density at the earth station receiver input (W/Hz)		R_c	transmission bit rate (b/s)
			R_{cinb}	transmission bit rate on inbound carrier (b/s)
OBO	output back-off		R_{coutb}	transmission bit rate on outbound carrier (b/s)
OBO_1	output back-off per carrier with multicarrier operation mode		R_e	earth radius: $R_e = 6378$ km
OBO_i	output back-off for interfering carrier		R_0	geostationary satellite altitude: $R_0 = 35\,786$ km
OBO_t	total output back-off with multicarrier operation mode		R_w	slant range from earth station to wanted satellite
OBO_w	output back-off for wanted carrier		S	normalised throughput
			SKW	satellite station keeping window halfwidth (degrees)
P	power (W)			
P_f	probability for a frame to be in error		T	interval of time (s), also period of orbit (s), also medium temperature (K) and noise temperature (K)
P_R	received power at antenna output (W)			
P_T	power fed to transmitting antenna (W)		T_A	antenna noise temperature (K)
P_{TX}	transmitter output power (W)		T_D	downlink system noise temperature (K)
P_{TXmax}	transmitter output power at saturation (W)		T_{Dmin}	minimum value of T_D (K)

T_F	feeder temperature (K)	η	efficiency
T_{GROUND}	ground noise temperature in vicinity of earth station (K)	η_a	antenna efficiency (typically 0.6)
		η_c	channel efficiency
T_{IF}	intermediate frequency amplifier effective input noise temperature (K)	η_{cGBN}	channel efficiency with go-back-N protocol
T_{LNA}	low noise amplifier effective input noise temperature (K)	η_{cSR}	channel efficiency with selective-repeat protocol
T_m	average medium temperature (K)	η_{cSW}	channel efficiency with stop-and-wait protocol
T_{MX}	mixer effective input noise temperature (K)	θ	angle from boresigth of antenna (degrees)
T_p	propagation time (s)	θ_{3dB}	half power beamwidth of an antenna (degrees)
T_R	receiver effective input noise temperature (K)	θ_R	antenna off-axis of angle for reception (degrees)
T_{RT}	round trip time (s)	θ_{Rmax}	maximum value of antenna off-axis angle for reception (degrees)
T_{SKY}	clear sky noise temperature (K)		
T_U	uplink system noise temperature (K)	θ_T	antenna off-axis angle for transmission (degrees)
THRU	throughput (b/s)	θ_{Tmax}	maximum value of antenna off-axis angle for transmission (degrees)
W	window size		
X	order of an intermodulation product	λ	wavelength (m) $= c/f$, also traffic generation rate (s^{-1})
XPD	cross polar discrimination	μ	product of gravitational constant G and mass of the Earth M_e: $\mu = 3.986 \times 10^{14}\,m^3/s^2$
XPI_{RX}	receive antenna cross polarisation isolation		
XPI_{TX}	transmit antenna cross polarisation isolation	ρ	code rate
α	angular separation between two satellites (degrees)	τ	packet duration (s)
		Φ	power flux density (W/m^2)
Γ	spectral efficiency (b/s Hz)	Φ_{sat}	power flux density at saturation (W/m^2)
Δ	ratio of co-polar wanted carrier power to cross-polar interfering carrier power	Φ_t	total flux density (W/m^2)

1 INTRODUCTION

This chapter aims at providing the framework of VSAT technology in the evolving context of satellite communications in terms of network configuration, services, operational and regulatory aspects. It also can be considered by the reader as a guide to the following chapters which aim at providing more details on the most important issues.

1.1   VSAT NETWORK DEFINITION

VSAT, now a well established acronym for Very Small Aperture Terminal, was initially a trade mark for a small earth station marketed in the 80s by Telcom General in the USA. Its success as a generic name probably comes from the appealing association of its first letter V, which establishes a 'victorious' context, or may be perceived as a friendly sign of participation, and SAT which definitely establishes a connection with satellite communications.

In this book, the use of the word 'terminal' which appears in the clarification of the acronym will be replaced by 'earth station', or station for short, which is the more common designation in the field of satellite communications for the equipment assembly allowing reception from or transmission to a satellite. The word terminal will be used to designate the end user equipment (telephone set, facsimile machine, television set, computer, etc.) which generates or accepts the traffic that is conveyed within VSAT networks. This complies with regulatory texts, such as those of the International Telecommunications Union (ITU), where for instance equipment generating data traffic, such as computers, is named 'Data Terminal Equipment' (DTE).

VSATs are one of the intermediate steps of the general trend in earth station size reduction that has been observed in satellite communications since the launch of the first communication satellites in the mid 60s. Indeed earth stations have evolved from the large INTELSAT Standard A earth stations equipped with antennas 30 m wide, to today's receive-only stations with antennas as small as 60 cm for direct reception of television transmitted by broadcasting satellites, or hand held terminals for radiolocation such as the Global Positioning System (GPS) receivers. Prospects are for telephone handsets intended for satellite personal communications [MAR94].

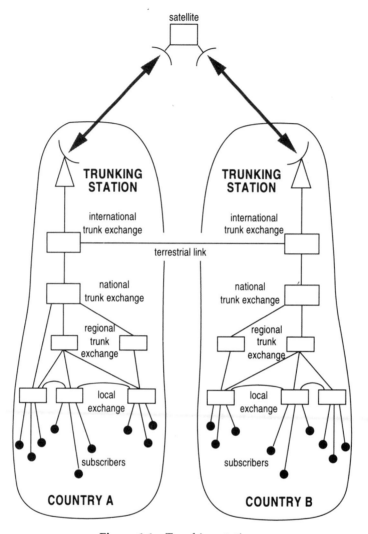

Figure 1.1   Trunking stations

Therefore VSATs are at the lower end of a product line which offers a large variety of communication services: at the upper end are large stations which support large capacity satellite links. They are mainly used within international switching networks to support trunk telephony services between countries, possibly on different continents. Figure 1.1 illustrates how such stations collect traffic from end users via terrestrial links that are part of the public switched network of a given country. These stations are quite expensive, with costs in the range of $10 million, and require important civil works for their installation. Link capacities are in the range of a few thousand telephone channels, or equivalently about 100 Mb/s. They are owned and operated by national telecom operators, such as the PTTs.

At the lower end are VSATs: these are small stations with antenna diameter less than 2.4 m, hence the name 'small aperture' which refers to the area of the antenna. Such stations cannot support satellite links with large capacities, but they are cheap, with manufacturing costs in the range of $5000 to $10 000, and easy to install any place, on the roof of a building or on a parking lot. Installation costs do not exceed $2000. Therefore, they are within the financial capabilities of small corporate companies, and can be used to set up rapidly small capacity satellite links, in a flexible way. Capacities are of the order of a few tens of kb/s, typically 56 or 64 kb/s.

Referring to transportation, VSATs are for information transport, the equivalent of personal cars for human transport, while the large earth stations mentioned earlier are like public buses or trains.

At this point it is worth noting that VSATs, like personal cars, are available at one's premises. This avoids the need for using any public network links to access

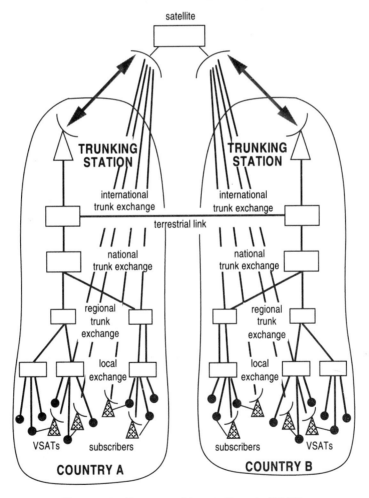

Figure 1.2   From trunking stations to VSATs

the earth station. Indeed the user can directly plug into the VSAT equipment his own communication terminals such as telephone or video set, personal computer, printer, etc. Therefore VSATs appear as natural means to by-pass public network operators by directly accessing satellite capacity. They are flexible tools for establishing private networks, for instance between the different sites of a company.

Figure 1.2 illustrates this aspect by emphasising the positioning of VSATs near the user compared to trunking stations, which are located at the top level of the switching hierarchy of a switched public network.

The by-pass opportunity offered by VSAT networks has not always been well accepted by national telecom operators as it could mean loss of revenues, as a result of business traffic being diverted from the public network. This has initiated conservative policies by national telecom operators opposed to the deregulation of the communications sector. In some regions of the world, and particularly in Europe, this has been a strong restraint to the development of VSAT networks.

1.2   VSAT NETWORK CONFIGURATIONS

As illustrated in Figure 1.2, VSATs are connected by radio frequency links via a satellite. Those links are radio frequency links, with a so-called 'uplink' from the station to the satellite and a so-called 'downlink' from the satellite to the station (Figure 1.3). The overall link from station to station, sometimes called hop, consists of an uplink and a downlink. A radio frequency link is a modulated carrier conveying information. Basically the satellite receives the uplinked carriers from the transmitting earth stations within the field of view of its receiving antenna, amplifies those carriers, translates their frequency to a lower band in order to avoid possible output/input interference, and transmits the amplified carriers to the stations located within the field of view of its transmitting antenna. A more detailed description of the satellite architecture is given in Chapter 2 (section 2.1).

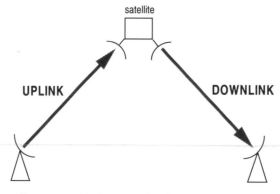

Figure 1.3   Definition of uplink and downlink

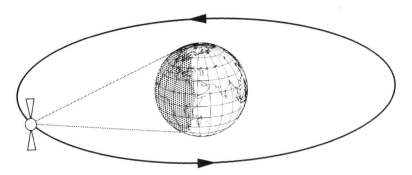

Figure 1.4 Geostationary satellite

Present VSAT networks use geostationary satellites, which are satellites orbiting in the equatorial plane of the earth at an altitude above the earth surface of 35786 km. It will be shown in Chapter 2 that the orbit period at this altitude is equal to that of the rotation of the Earth. As the satellite moves on its circular orbit in the same direction as the earth rotates, the satellite appears from any station on the ground as a fixed relay in the sky. Figure 1.4 illustrates this geometry. It should be noted that the distance from an earth station to the geostationary satellite induces a radio frequency carrier power attenuation of typically 200 dB on both uplink and downlink and a propagation delay from earth station to earth station (hop delay) of about 0.25 s (see Chapter 2).

As a result of its apparent fixed position in the sky, the satellite can be used 24 hours a day as a permanent relay for the uplinked radio frequency carriers. Those carriers are downlinked to all earth stations visible from the satellite (shaded area on the earth in Figure 1.4). Thanks to its apparent fixed position in the sky, there is no need for tracking the satellite. This simplifies VSAT equipment and installation.

As all VSATs are visible from the satellite, carriers can be relayed by the satellite from any VSAT to any other VSAT in the network, as illustrated by Figure 1.5.

Regarding meshed VSAT networks, one must take into account the following limitations:

—typically 200 dB carrier power attenuation on the uplink and the downlink as a result of the distance to and from a geostationary satellite;

—limited satellite radio frequency power, typically a few tens of watts;

—small size of the VSAT, which limits its transmitted power and its receiving sensitivity.

As a result of the above, it may well be that the demodulated signals at the receiving VSAT do not match the quality requested by the user terminals. Therefore direct links from VSAT to VSAT may not be acceptable.

The solution then is to install in the network a station larger than a VSAT, called the *hub*. The hub station has a larger antenna size than that of a VSAT, say 4 m to

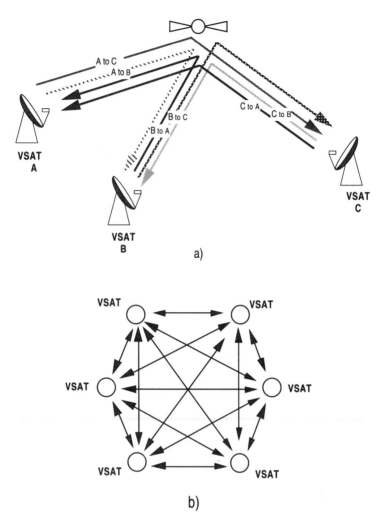

Figure 1.5 Meshed VSAT network. (a) Example with three VSATs (arrows represent information flow as conveyed by the carriers relayed by the satellite). (b) Simplified representation for a larger number of VSATs (arrows represent bi-directional links made of two carriers travelling in opposite directions)

11 m, resulting in a higher gain than that of a typical VSAT antenna, and is equipped with a more powerful transmitter. As a result of its improved capability, the hub station is able to receive adequately all carriers transmitted by the VSATs, and to convey the desired information to all VSATs by means of its own transmitted carriers. The architecture of the network becomes star shaped as shown in Figures 1.6 and 1.7. The links from the hub to the VSAT are named 'outbound links'. The ones from the VSAT to the hub are named 'inbound links'. Both inbound and outbound links consist of two links, uplink and downlink, to and from the satellite, as illustrated in Figure 1.3.

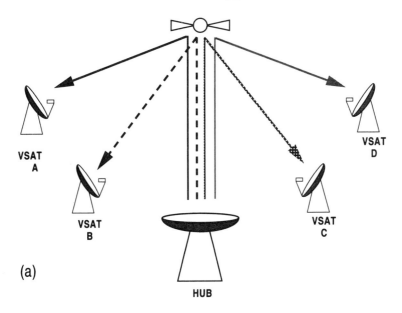

(a)

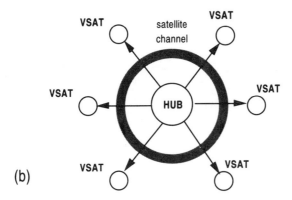

(b)

Figure 1.6   One-way star-shaped VSAT network. (a) Example with four VSATs (arrows represent information flow as conveyed by the outbound carriers relayed by the satellite). (b) Simplified representation for a larger number of VSATs (arrows represent unidirectionnal links)

There are two alternatives to star shaped VSAT networks:

—One-way networks (Figure 1.6), where the hub transmits carriers to receive-only VSATs. This configuration supports broadcasting services from a central site where the hub is located to remote sites where the receive-only VSATs are installed.

—Two-way networks (Figure 1.7), where VSATs can transmit and receive. Such networks support interactive traffic.

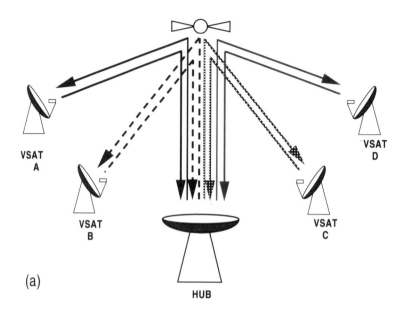

(a)

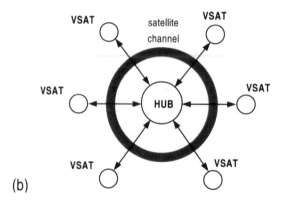

(b)

Figure 1.7   Two-way star shaped VSAT network. (a) Example with four VSATs (arrows represent information flow as conveyed by the carriers relayed by the satellite). (b) Simplified representation for a larger number of VSATs (arrows represent bi-directional links made of two carriers travelling in opposite directions)

The two-way connectivity between VSATs can be achieved in two ways:

—either direct links from VSAT to VSAT via satellite, should the link performance meet the requested quality (this is the mesh configuration illustrated in
Figure 1.5),

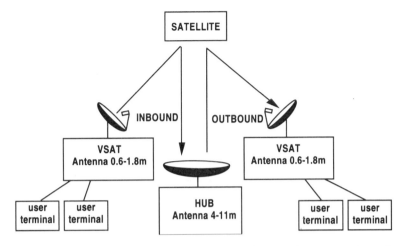

Figure 1.8 VSAT to VSAT connectivity using the hub as a relay in star shaped networks

—or by double hop links via satellite in a star shaped network, with a first hop from VSAT to hub and then a second hop using the hub as a relay to the destination VSAT (Figure 1.8).

In conclusion, star shaped networks are imposed by power requirements resulting from the reduced size and hence the low cost of the VSAT earth station in conjunction with power limitation of the satellite. Meshed networks are considered whenever such limitations do not hold, or are unacceptable. Meshed networks have the advantage of a reduced propagation delay (single hop delay is 0.25 s instead of 0.5 s for double hop) which is especially of interest for telephony service.

1.3 VSAT NETWORK APPLICATIONS AND TYPES OF TRAFFIC

VSAT networks have both civilian and military applications. These will now be presented.

1.3.1 Civilian VSAT networks

1.3.1.1 Types of services

As mentioned in the previous section, VSAT networks can be configured as one-way or two-way networks. Table 1.1 gives examples of services supported by VSAT networks according to these two classes.

Table 1.1   Examples of services supported by VSAT networks

One-way VSAT networks
   Stock market and other news broadcasting
   Training or continuing education at distance
   Distribute financial trends and analyses
   Introduce new products at geographically dispersed locations
   Update market related data, news and catalogue prices
   Distribute video or TV programmes
   Distribute music in stores and public areas
   Relay advertising to electronic signs in retail stores

Two-way VSAT networks
   Interactive computer transactions
   Low rate video conferencing
   Database enquiries
   Bank transactions, automatic teller machines
   Reservation systems
   Distributed remote process control and telemetry
   Voice communications
   Emergency services
   Electronic funds transfer at point of sale
   E-mail
   Medical data transfer
   Sales monitoring and stock control
   Satellite news gathering

It can be noticed that most of the services supported by two-way VSAT networks deal with interactive data traffic, where the user terminals are most often personal computers. The most notable exceptions are voice communications and satellite news gathering.

Voice communications on a VSAT network means telephony with possibly longer delays than that incurred on terrestrial lines as a result of the long satellite path. Telephony services imply full connectivity, and delays are typically 0.25 s or 0.50 s depending on the selected network configuration, as mentioned above.

Satellite News Gathering (SNG) can be viewed as a temporary network using transportable VSATs, sometimes called 'fly-away' stations, which are transported by car or aircraft and set up at a location where news reporters transmit video signals to a hub located near the company's studio. Of course the service could be considered as inbound only, if it were not for the need to check the uplink from the remote site, and to be in touch by telephone with the staff at the studio. As 'fly-away' VSATs are constantly transported, assembled and disassembled, they must be robust, lightweight and easy to install. Today they weigh typically 250 kg and can be installed in 20 minutes. Figure 1.9 shows a picture of a 'fly away' VSAT station [ELI93].

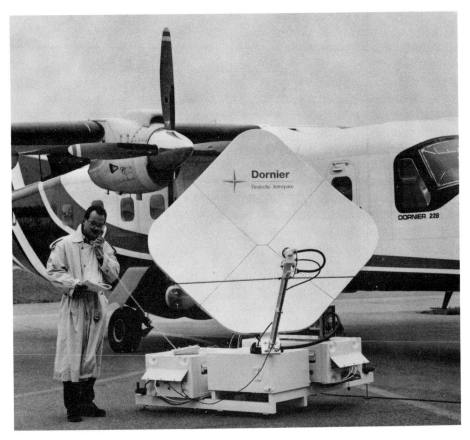

Figure 1.9 'Fly away' VSAT station. (Reproduced from [ELI93] by permission of the Institution of Electrical Engineers, © 1993 IEE)

1.3.1.2   Types of traffic

Depending on the service the traffic flow between the hub and the VSATs may have different characteristics and requirements.

Data transfer or broadcasting, which belongs to the category of one-way services, typically displays file tranfers of 1–100 Mbytes of data. This kind of service is not delay sensitive, but requires a high integrity of the data which are transferred. Examples of applications are computer download and distribution of data to remote sites.

Interactive data is a two-way service corresponding to several transactions per minute and per terminal of single packets 50 to 250 bytes long on both inbound and outbound links. The required response time is typically a few seconds. Examples of applications are bank transactions, and electronic funds transfer at point of sale.

Table 1.2 Types of traffic

Type of traffic	Packet length		Required response time	Usage mode	Examples
	Inbound	Outbound			
Data transfer or broadcasting	Not relevant	1–100 Mbytes	Not delay sensitive	—	Computer down load, distribution of data to remote sites
Interactive data	50–250 bytes	50–250 bytes	A few seconds	Several transactions per minute per terminal	Bank transactions, electronic funds transfer at point of sale
Enquiry/response	30–100 bytes	500–2000 bytes	A few seconds	Several transactions per minute per terminal	Airline reservations, database enquiries
Supervisory control and data acquisition (SCADA)	100 bytes	10 bytes	A few seconds/minutes	One transaction per second/minute per terminal	Control/monitoring of pipeline and off-shore platforms, electric utilities and water resources

Enquiry/response is a two-way service corresponding to several transactions per minute and terminal. Inbound packets (typically 30–100 bytes) are shorter than outbound packets (typically 500–2000 bytes). The required response time is typically a few seconds. Examples of applications are airline or hotel reservations and database enquiries.

Supervisory control and data acquisition (SCADA) is a two-way service corresponding to one transaction per second or minute per terminal. Inbound packets (typically 100 bytes) are longer than outbound packets (typically 10 bytes). The required response time ranges from a few seconds to a few minutes. What is most important is the high data security level, and the low power consumption of the terminal. Examples of applications are control and monitoring of pipelines, off-shore platforms, electric utilities and water resources.

Table 1.2 summarises the above discussion.

1.3.2 Military VSAT networks

VSAT networks have been adopted by many military forces in the world. Indeed the inherent flexibility in the deployment of VSATs makes them a valuable means to install temporary communications links between small units in the battlefield and headquarters located near the hub [WEL93]. Moreover the topology of a star shaped network fits well into the natural information flow between field units and command base. Frequency bands are at X-band, with uplinks in the 7.9–8.4 GHz band and downlinks in the 7.25–7.75 GHz band.

The military user VSAT must be a small, low weight, low power station that is easy to operate under battlefield conditions. As an example, the manpack station developed by the UK Defence Research Agency (DRA) for its Milpico VSAT military network is equipped with a 45 cm antenna, weighs less than 17 kg and can be set up within 90 seconds [WEL93]. It supports data and vocoded voice at 2.4 kb/s. In order to do so, the hub stations need to be equipped with antennas as large as 14 m. Another key requirement is low probability of detection by hostile interceptors. Spread spectrum techniques are largely used [EVA90, Chapter 15].

1.4 VSAT NETWORKS: INVOLVED PARTIES

The applications of VSAT networks identified in the previous section clearly indicate that VSAT technology is appropriate to business or military applications. Reasons are the inherent flexibility of VSAT technology, as mentioned above in section 1.1, cost savings and reliability, as will be discussed in section 3.3.

Which are the involved parties, as far as corporate communications are concerned?

—The *user* is most often a company employee using office communication *terminals* such as personal computers, telephone sets, and fax machines. On other occasions the terminal is transportable, as with satellite news gathering (SNG). Here the user is mostly interested in transmitting video to the company studio. The terminal may be fixed but not located in an office as with SCADA (supervisory control and data acquisition) applications.

—The VSAT *network operator* may be the user's company itself, if the company owns the network, or it may be a telecom company (in many countries it is the national public telecom operator) who then leases the service. The VSAT network operator is then a customer to the network and/or the equipment provider.

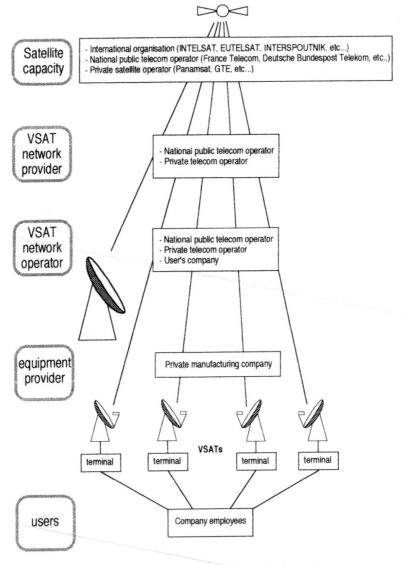

Figure 1.10   VSAT networks: involved parties

—The VSAT *network provider* has the technical ability to dimension and install the network. It elaborates the network management system (NMS) and designs the corresponding software. Its inputs are the customer's needs, and its customers are network operators. The network provider may be a private company or a national telecom operator.

—The *equipment provider* sells the VSATs and/or the hub which it manufactures. It may be the network provider or a different party.

For the VSAT network to work, some *satellite capacity* must be provided. The satellite may be owned by the user's company but this is a rare example of 'vertical integration', and most often the satellite is operated by a different party. This party may be a private satellite operator, a national satellite operator, or an international organisation such as INTELSAT or EUTELSAT. In the latter case, and up till now, only a signatory to the organisation is allowed to lease satellite capacity and to provide it secondhand. Most often the signatory is a national telecom operator.

The above parties are those involved in the contractual matters. Other parties are on the regulatory side and their involvement will be first presented in section 1.9 and developed with more details in Chapter 3.

Figure 1.10 summarises the above discussion. The terminology will be used throughout the book, and therefore Figure 1.10 can serve as a convenient reference.

1.5   VSAT NETWORK OPTIONS

1.5.1   Star or mesh?

Section 1.2 introduced the two main architectures of a VSAT network: star or mesh. The question now is: is one architecture more appropriate than the other?

The answer depends on three factors:

—the structure of information flow within the network;

—the requested link quality and capacity;

—the transmission delay.

These three aspects will now be discussed.

1.5.1.1   *Structure of information flow*

VSAT networks can support different types of applications, and each has an optimum network configuration:

Table 1.3   VSAT network configuration appropriate to a specific application

Application	Network configuration		
	Star shaped one-way	Star shaped two-way	Meshed two-way
Broadcasting	×		
Corporate network (hub at company headquarters, VSATs at branches)		×	
Corporate network (distributed sites)		× (double hop)	× (single hop)

—Broadcasting: a central site distributes information to many remote sites with no back flow of information. Hence a star shaped one-way network supports the service at the lowest cost.

—Corporate network: most often companies have a centralised structure with administration and management performed at a central site, and manufacturing or sales performed at sites scattered over a geographical area. Information from the remote sites needs to be gathered at the central site for decision making, and information from the central site has to be distributed to the remote ones, such as task sharing. Such an information flow can be partially supported by a star shaped one-way VSAT network, for instance for information distribution, or totally supported by a two-way star shaped VSAT network. In the first case, VSATs need to be receive-only and are less expensive than in the latter case where interactivity is required, as this implies VSATs equipped with both transmit and receive equipments. Typically the cost of the transmitting equipment is two-thirds that of an interactive VSAT.

—interactivity between distributed sites: other companies or organisations with a decentralised structure are more likely to comprise many sites interacting one with another. A meshed VSAT network using direct single hop connections from VSAT to VSAT is hence mostly desirable. The other option is a two-way star shaped network with double hop connections from VSAT to VSAT via the hub.

Table 1.3 summarises the above discussion. Regulatory aspects also have to be taken into account (see Chapter 3).

1.5.1.2   Link quality and capacity

The link considered here is the link from the transmitting station to the receiving one. Such a link may comprise several parts: for instance a single hop link would

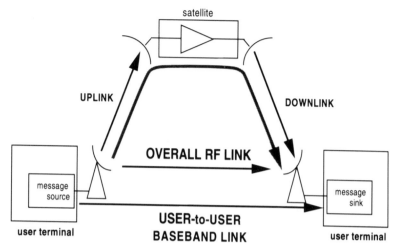

Figure 1.11 Overall radio frequency (RF) link and user-to-user baseband link

comprise an uplink and a downlink (Figure 1.3), a double hop link would comprise two single hop links, one being inbound and the other outbound (Figure 1.8).

When dealing with link quality, one must refer to the quality of a given signal. Actually two types of signals are involved: the modulated carrier at the input to the receiver and the baseband signals delivered to the user terminal once the carrier has been demodulated (Figure 1.11). The input to the receiver terminates the *overall radio frequency link* from the transmitting station to the receiving one, with its two link components, the uplink and the downlink. The earth station interface to the user terminal terminates the *user-to-user baseband link* from the output of the device generating bits (message source) to the input of the device to which those bits are transmitted (message sink).

The link quality of the radio frequency link is measured by the $(C/N_0)_T$ ratio at the station receiver input, where C is the received carrier power and N_0 the power spectral density of noise [MAR93, Chapter 2].

The baseband link quality is measured by the information bit error rate (BER). It is conditioned by the E_b/N_0 value at the receiver input, where E_b (J) is the energy per information bit and N_0 (W/Hz) is the noise power spectral density. As indicated in Chapter 5, section 5.7, the E_b/N_0 ratio depends on the overall radio frequency link quality $(C/N_0)_T$ and the capacity of the link, measured by its information bit rate R_b (b/s):

$$\frac{E_b}{N_0} = \frac{(C/N_0)_T}{R_b} \tag{1.1}$$

Figure 1.12 indicates the general trend which relates EIRP to *G/T* in a VSAT network, considering a given baseband signal quality in terms of constant BER. EIRP designates the effective isotropic radiated power of the transmitting

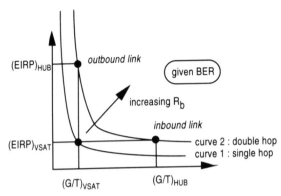

Figure 1.12   EIRP versus *G/T* in a VSAT network. Curve 1: single hop from VSAT to VSAT in a meshed network. Curve 2: double hop from VSAT to VSAT via the hub

equipment and G/T is the figure of merit of the receiving equipment (see Chapter 5, for definition of the EIRP and of the figure of merit).

As can be viewed from Figure 1.12, the double hop from VSAT to VSAT via the hub, when compared to a single hop, allows an increased link capacity, without modifying the size of the VSATs. Now this option also involves a larger transmission delay.

1.5.1.3   Transmission delay

With a single hop link from VSAT to VSAT in a meshed network, the propagation delay is about 0.25 s. With a double hop from VSAT to VSAT via the hub, the propagation delay is twice as much, i.e. about 0.5 s.

Double hop may be a problem for voice communications. However, it is not a severe problem for video or data transmission.

Table 1.4 summarises the above discussion: given the EIRP and G/T values for a VSAT, the designer can decide for either a large delay from VSAT to VSAT and

Table 1.4   Characteristics of star and mesh network configuration.

	Network configuration	
	Star shaped (double hop)	Mesh (single hop)
Capacity (given VSAT EIRP and G/T)	large	small
Delay (from VSAT to VSAT)	0.5 s	0.25 s

a larger capacity or a small delay and a lower capacity, by implementing either a star network, or a mesh one.

1.5.2 Data/voice/video

Depending on his needs, the customer may want to transmit either one kind of signal, or a mix of different signals. Data and voice are transmitted in a digital format, while video may be analogue or digital. When digital, the video signal may benefit from bandwidth efficient compression techniques.

1.5.2.1 Data transmission

VSATs have emerged from the need to transmit data. Standard VSAT products offer data transmission facilities. Rates offered to the user range typically from 50 b/s to 64 kb/s with interface ports such as RS-232, V24 and V28 for bit rates lower than 20 kb/s, and RS-422, RS-449, V11, V35 and X21 for higher bit rates. Appendix 3 gives some details on the functions of such ports.

Data distribution can be implemented in combination with video transmission using the Multiplexed Analogue Components (MAC) standard (see below: Video transmission). MAC also allows data transmission only (data occupy the full video frame and then no video is transmitted). Capacity then is as high as 20 Mb/s.

1.5.2.2 Voice transmission

Voice communications are of interest on two-way networks only. They can be performed at low rate using voice encoding (vocoder). Typical information rate then ranges from 4.8 kb/s to 9.6 kb/s. They can also be combined with data transmission (for instance up to four voice channels may be multiplexed with data or facsimile channels on a single 64 kb/s channel). On VSAT networks voice communications suffer from delay associated with vocoder processing (about 50 ms) and propagation on satellite links (about 500 ms for a double hop). Therefore the user may prefer to connect to terrestrial networks which offer a reduced delay. Voice communications can be a niche market for VSATs as a service to locations where land lines are not available, or for transportable terminal applications.

1.5.2.3 Video transmission

On the outbound link (from hub to VSAT), video transmission makes use of usual TV standards (NTSC, PAL or SECAM) in combination with FM modulation, or can be implemented on Multiplexed Analogue Components (MAC) standards (B MAC or D2 MAC), possibly in combination with distribution of data.

On the inbound link, as a result of the limited power of the VSAT on the uplink, video transmission is feasible at a low rate, possibly in the form of slow motion image transmission using video coding and compression.

1.5.3   Fixed/demand assignment

The earth stations of a VSAT network communicate via the satellite by means of modulated carriers. Any such carrier is assigned a portion of the resource offered by the satellite in terms of powered bandwidth. This assignment can be defined once for all, and this is called 'fixed assignment' (FA), or in accordance with requests from the VSATs depending on the traffic they have to transmit, and this is called 'demand assignment' (DA).

1.5.3.1   *Fixed assignment (FA)*

Figure 1.13 illustrates the principle of *fixed assignment*. A star shaped network configuration is considered in the figure but the principle applies to a meshed network configuration as well. The satellite resource is shared in a fixed manner by all stations whatever the traffic demand. It may be that at a given instant the VSAT traffic load is larger than that which can be accommodated by capacity allocated to that VSAT as determined by its share of the satellite resource. The VSAT must store or reject the traffic demand, and this either increases the delay, or introduces blocking of calls, in spite of the fact that other VSATs may have excess capacity available. Because of this, the network is not optimally exploited.

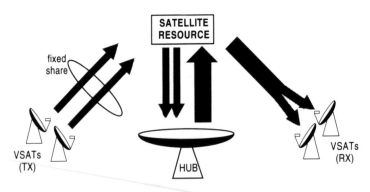

Figure 1.13   Principle of fixed assignment

1.5.3.2   Demand assignment (DA)

With *demand assignment*, VSATs share a variable portion of the overall satellite resource as illustrated in Figure 1.14. VSATs use only the capacity which is required for their own transmission, and leave the capacity in excess for use by other VSATs. Of course this variable share can be exercised only within the limits of the total satellite capacity allocated to the network.

Demand assignment is performed by means of requests for capacity transmitted by individual VSATs. Those requests are transmitted to the hub station, or to a traffic control station, should the management of the demand assignment technique be centralised, or to all other VSATs, if the demand assignment is distributed. Those requests are transmitted on a specific signalling channel, or piggy-backed on the traffic messages. With centralised management, the hub station or the traffic control station replies by allocating to the VSAT the appropriate resource, either a frequency band or a time slot. With distributed management, all VSATs keep a record of occupied and available resource. This is discussed in more detail in Chapter 4, section 4.6.

From the above, it can be recognised that demand assignment offers a better use of the satellite resource but at the expense of a higher system cost and a delay in connection set-up. However, a larger number of stations can share the satellite resource. Hence the higher investment cost is compensated for by a larger return on investment.

The centralised/distributed management option depends on the network architecture: a centralised control is easier to perform with a star shaped network, as all traffic flows through the hub, which then is the natural candidate for demand assignment control. With a mesh shaped network, both centralised and distributed control can be envisaged. Delay for link set-up is shorter with distributed control, as a single hop (about 0.25 s) is sufficient to inform all VSATs in the network of the request and the corresponding resource occupancy, while a double hop (about 0.5 s) is necessary for the request to proceed to the central station, and for that station to allocate the corresponding resource. Finally, as

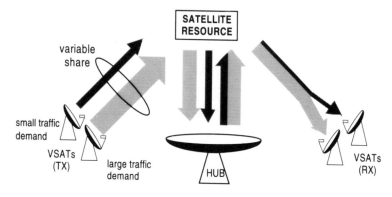

Figure 1.14   Principle of demand assignment

demand assignment implies charging the remote sites according to the resource occupancy, billing and accounting is more easily handled by a centralised control.

1.5.4   C-band or Ku-band?

VSAT networks are supposed to operate within the so-called 'fixed satellite service' (FSS) defined within the International Telecommunication Union (ITU).

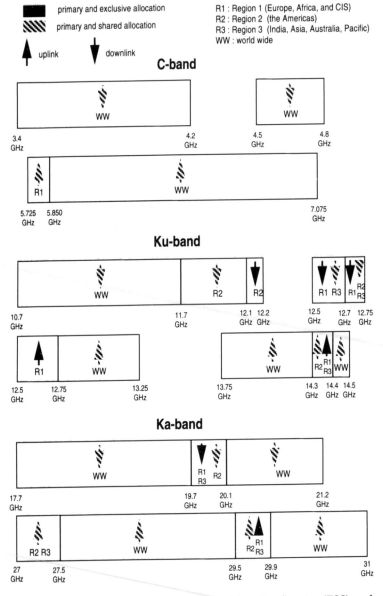

Figure 1.15   Frequency bands allocated to the Fixed Satellite Service (FSS) and usable for VSAT networks [ITU90]

The only exception is when data are broadcast in association with broadcasting of television or audio programmes, within the so-called 'broadcasting satellite service' (BSS).

The FSS covers all satellite communications between stations located while operating at given 'specified fixed points' of the Earth. Transportable stations belong to this category, and hence the so-called 'fly-away' stations should use the same frequency bands as fixed VSATs.

The most commonly used bands for commercial applications are those allocated to the FSS at C-band and Ku-band, as indicated in Figure 1.15.

The figure displays uplinks and downlinks by means of arrows oriented upward or downward. The black arrows indicate a primary and exclusive allocation for FSS, which means in short that the FSS is protected against interference from any other service, which is then considered as secondary. The striped arrows indicate a primary but shared allocation, which means that the allocated frequency bands can also be used by services other than FSS with the same rights. Coordination is then mandatory, according to the procedure described in the ITU Radio Regulations.

The FSS also has bands allocated at X-band (about 8 GHz uplink and 7 GHz downlink) and Ka-band (about 14 GHz uplink and 12 GHz downlink). X-band is occupied by military systems and Ka-band is at present used by experimental systems only.

As mentioned above, data may be carried in association with video signals within the frequency band allocated to the broadcasting satellite service. Possible bands are 11.7–12.5 GHz in regions 1 and 3, and 12.2–12.7 GHz in region 2, filling in the gaps of the bands represented in Figure 1.15 which deals with the fixed satellite service only.

The selection of a frequency band for operating a VSAT network depends first on the availability of satellites covering the region where the VSAT network is to be installed. C-band satellites are available in most regions of the world (actually only the high latitudes above about $70°$ are not covered) while Ku-band satellites are available mainly over North America, Europe, East Asia and Australia. Figure 1.16 gives a general picture of the regions of the world where C-band or Ku-band satellite coverage is available.

To be considered next is the potential problem of interference.

Interference designates unwanted carriers entering in the receiving equipment along with the wanted ones. The unwanted carriers perturb the demodulator by acting as noise adding to the natural thermal noise. Interference is a problem with VSATs because the small size of the antenna (small aperture) translates into a radiation pattern with a large beamwidth. Indeed as shown by equation (1.2) the half power beamwidth θ_{3dB} of an antenna relates to the product of its diameter by frequency (see Appendix 4), as follows:

$$\theta_{3dB} = 70 \frac{c}{Df} \quad \text{(degrees)} \tag{1.2}$$

where $D(m)$ is the diameter of the antenna, $f(Hz)$ is the frequency, and $c = 3 \times 10^8 \, m/s$ is the velocity of light.

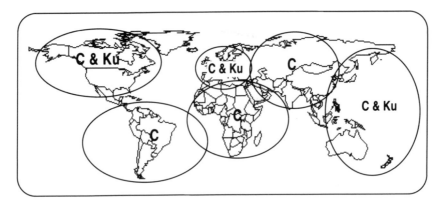

Figure 1.16 Regions of the earth where C-band and Ku-band satellite coverage is available

Therefore, the smaller the antenna diameter, the larger the beamwidth, and the off-axis interfering carriers are more likely to be emitted or received with high antenna gain. How important this perturbation can be is discussed in Chapter 5, section 5.5.

At this point it suffices to mention that interference is more likely to be a problem at C-band than at higher frequencies. There are two reasons: first, there is no primary and exclusive allocation to FSS at C-band. Second, given the earth station antenna diameter, interference is more important at C-band than at Ku-band, as the beamwidth is inversely proportional to the frequency, and thus is larger at C-band than at higher frequencies. To put this in perspective, formula (1.2) indicates for a 1.8 m antenna a beamwidth angle of 3° at 4 GHz, and only 1° at 12 GHz. This means that the receiving antenna is more likely to pick up carriers downlinked from satellites adjacent to the wanted one at C-band than at Ku-band, especially as C-band satellites are many and hence nearer each other. A typical angular separation for C-band satellites is 3°, and is therefore comparable to VSAT antenna beamwidth.

The same problem occurs on the uplink where a small VSAT antenna projects carrier power in a larger angle at C-band than at Ku-band, and hence generates more interference on the uplink of adjacent satellite systems. However this is not a major issue as the transmit power of VSATs is weak.

Finally it should be known that C-band and parts of Ku-band are shared by terrestrial microwave relays, and this may be another source of interference. Ku-band offers dedicated bands free from any terrestrial microwave transmission (see black arrows in Figure 1.15), which is not the case for C-band. This simplifies the positioning of the VSAT and hub station as no coordination is implied.

Figure 1.17 summarises the various interfering paths mentioned above.

Where the small size of the antenna is at a premium, and should interference be too large, interference can be combated by using a modulation technique named

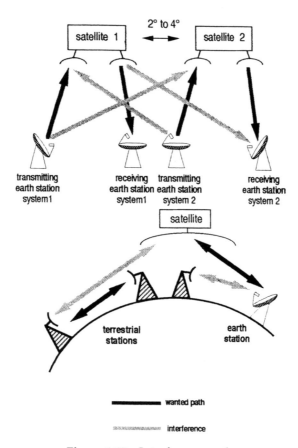

Figure 1.17 Interference paths

spread spectrum which consists in spreading the carrier in a much larger bandwidth than strictly required to transmit the information. This is an interesting technique as it provides not only interference protection but also potential for code division multiple access (CDMA) to a satellite channel, as will be shown in Chapter 4, section 4.6. However, as a result of the higher utilised bandwidth, it is less bandwidth efficient compared to alternative multiple access techniques such as Frequency Division Multiple Access (FDMA) or Time Division Multiple Access (TDMA), which can be used where interference is not too severe.

Finally, the cost of the equipment is another driving factor for selecting between C-band and Ku-band. Although C-band technology is cheaper, the larger size of the VSAT antenna for a similar performance makes the VSAT more expensive than at Ku-band.

Table 1.5 summarises the advantages and drawbacks of the most commonly available frequency bands.

Table 1.5 Advantages and drawbacks of the most commonly available frequency bands

	Advantages	Drawbacks
C-band	Worldwide availability Cheaper technology Robust to rainfall	Larger stations (1 to 3 m) Severe interference from adjacent satellites and terrestrial microwave sharing same frequency bands (may impose use of spread spectrum techniques and CDMA)
Ku-band	Makes better use of satellite capacity (possible use of more efficient access schemes such as FDMA or TDMA compared to CDMA) Smaller stations (0.6 to 1.8 m)	Limited (regional) availability Rain (attenuation and to a lesser extent depolarisation) affects link performance

1.5.5   Hub options

1.5.5.1   *Dedicated hub*

A dedicated hub supports a full single network with possibly thousands of VSATs connected to it. The hub may be located at the customer's organisation central site, with the host computer directly connected to it. It offers the customer full control of the network. In periods of expansion, changes in the network, or problems, this option may simplify the customer's life. However, a dedicated hub represents the most expensive option and is only justified if its cost can be amortised over a sufficiently large number of VSATs in the network. The typical cost of a dedicated hub is in the range of $1 million.

1.5.5.2   *Shared hub*

Several separate networks may share a unique hub. With this option hub services are leased to VSAT network operators. Hence the network operators are faced with minimum capital investment and this favours the initial implementation of a VSAT network. Therefore shared hubs are most suitable for the smaller networks (less than 50 VSATs). However sharing a hub has a number of drawbacks:

Need for a connection from hub to host.   A shared hub facility is generally not colocated with the customer's host computer. Hence a backhaul circuit is needed to connect the hub to the host. The circuit may be a leased line or one provided by a terrestrial switched telephone network. This adds an extra cost to the VSAT network operation. Moreover, operational experience has shown the backhaul circuit to be the weakest link in the chain. Therefore this option means an increased failure risk. A possible way to mitigate this potential problem area would be using route diversity: for instance a microwave or satellite link could be used as a back-up for this interconnection.

Possible limitation in future expansion. A shared hub may impose an unforeseen capacity limitation, as the available capacity may be leased without notice to the other network operators sharing the hub. Guarantees should contractually be asked for by any network operator in this regard.

1.5.5.3 Mini-hub

The mini-hub is a small hub, with an antenna size as low as 2–3 m and a typical cost in the range of $100 000. It appeared rather recently (1989), as a result of the increased power from satellites and the improved performance of low noise receiving equipment. The mini-hub has been shown to be a very attractive solution, as it retains the advantages of a dedicated hub at a reduced cost. It also eases possible installation problems in connection with downtown areas or communities with zoning restrictions, as a mini-hub entails a smaller antenna size and less rack mounted equipment than a large dedicated hub or even a shared hub. A typical mini-hub can support 300 to 400 remote VSATs.

1.6   VSAT NETWORK EARTH STATIONS

1.6.1   VSAT station

Figure 1.18 illustrates the architecture of a VSAT station. As shown in the figure, a VSAT station is made of two separate equipments: the outdoor unit (ODU) and the indoor unit (IDU). The outdoor unit is the VSAT interface to the satellite, while the indoor unit is the interface to the customer's terminals or local area network (LAN).

1.6.1.1   The outdoor unit (ODU)

Figure 1.19 shows a photograph of an outdoor unit, with its antenna and the electronics package which contains the transmitting amplifier, the low noise receiver, the up- and down-converters and the frequency synthesiser. The photograph of Figure 1.20 provides a closer look at the electronics container.
    For a proper specification of the ODU, as an interface to the satellite, the following parameters are of importance:

—the transmit and receive frequency bands;

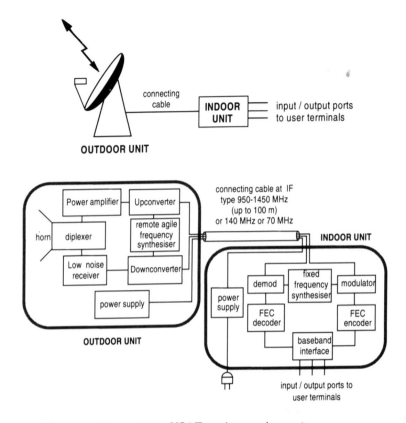

Figure 1.18   VSAT station equipments

—the transmit and receive step size for adjusting the frequency of the transmitted carrier or for tuning to the received carrier frequency;

—the effective isotropic radiated power (EIRP), which conditions the perform-ance of the radio frequency uplink. The EIRP depends on the value of the antenna gain, and hence its size and transmit frequency, and on the transmit-ting amplifier output power (see Chapter 5, section 5.2);

—the figure of merit G/T, which conditions the performance of the radio fre-quency downlink. The G/T ratio depends on the value of the antenna gain, and hence its size and receive frequency, and on the noise temperature of the receiver (see Chapter 5, section 5.3);

—the antenna side-lobe gain variation with off-axis angle which controls the off-axis EIRP and G/T, hence conditioning the levels of produced and received interference.

Figure 1.19 Photograph of the outdoor unit of a VSAT station. (Reproduced by permission of GTE Spacenet)

Operating temperature range, wind loading under operational and survival conditions, rain and humidity are also to be considered.

1.6.1.2 The indoor unit (IDU)

The indoor unit installed at the user's facility is shown on the photograph of Figure 1.21. In order to connect his terminals to the VSAT, the user must access the ports installed on the rear panel of the outdoor unit, shown on the photograph of Figure 1.22.

For a proper specification of the IDU, as an interface to the the user's terminals or to a local area network (LAN), the following parameters are of importance:

—number of ports;

Figure 1.20 Photograph of the electronics container of the outdoor unit of Figure 1.19. (Reproduced by permission of GTE Spacenet)

Figure 1.21 Front view of the indoor unit of a VSAT station. (Reproduced by permission of GTE Spacenet)

Figure 1.22 Rear view of the indoor unit of Figure 1.21. (Reproduced by permission of GTE Spacenet)

—type of ports: mechanical, electrical, functional and procedural interface. This is often specified by reference to a standard, such as those mentioned in the above section (VSAT network options) and in Appendix 3;

—port speed: this is the maximum bit rate at which data can be exchanged between the user terminal and the VSAT indoor unit on a given port. The actual data rate can be lower.

1.6.2   Hub station

Figure 1.23 shows a photograph of a hub station and Figure 1.24 displays the architecture of the hub station with its equipments. Apart from the size and the number of subsystems, there is little functional difference between a hub and a VSAT, so that most of the content of the above section applies here. The major difference is that the indoor unit of a hub station interfaces either to a host computer or to a public switched network or private lines, depending on whether the hub is a dedicated or a shared one (see the above section on VSAT network options). Typical hub station parameters are indicated in Table 1.6.

One can notice in Figure 1.24 that the hub station is equipped with a Network Management System (NMS). The NMS is a mini-computer or a workstation, used for configuration, control, alarm and traffic monitoring, operator interface and batch offline activities such as printing reports. The NMS is in charge of routeing procedures, monitoring and controlling the network status, performing resets

Figure 1.23 Photograph of the outdoor unit of a hub station. (Courtesy of Hughes Network Systems)

and diagnostics, enabling or disabling network components, adding or deleting VSAT terminals, network interfaces, and satellite channels. Software can be downloaded from the NMS to the remote VSATs. Finally, the NMS performs administrative duties such as maintaining an inventory of all equipment, recording the usage of the network and preparing the billing.

The above wide list of functions to be performed by the NMS shows its important role for the network. Actually, the adequacy of the NMS's response to the user's needs makes the difference between popular network providers and those who fail to survive in the market.

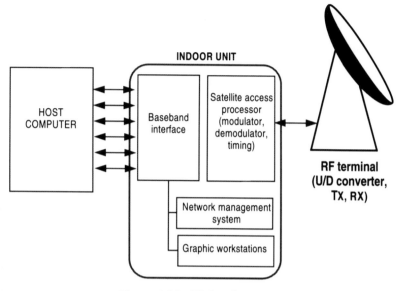

Figure 1.24 Hub subsystems

Table 1.6 Typical hub station parameters

	Compact hub	Medium hub	Large hub
Antenna diameter	2 to 5 m	5 to 8 m	8 to 10 m
Transmitter power:			
Ku-band	3–15 W SSPA*	3–15 W SSPA*	50–100 W TWT†
C-band	5–20 W SSPA	5–20 W SSPA	100–200 W TWT
Receiver noise temperature:			
Ku-band	80–120 K	80–120 K	80–120 K
C-band	35–55 K	35–55 K	35–55 K
Cost	about $100 000	about $500 000	about $1 million

*SSPA: Solid State Power Amplifier.
†TWT: Travelling Wave Tube.

1.7 HISTORICAL BACKGROUND

1.7.1 Origin of VSAT networks

In 1979, the Equatorial Communications Company started VSAT services offering initially one-way communications from a hub located at Mountain View, CA, to receive-only VSATs in the USA. These VSATs were known as C-100 type VSATs. Links were at C-band and used spread spectrum techniques to avoid interference from adjacent satellites on the downlink as a result of the wide beam of VSAT

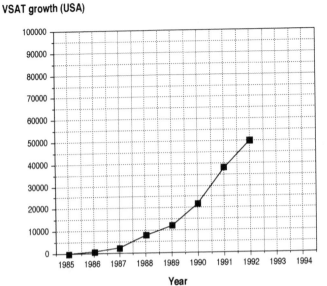

Figure 1.25   VSAT growth in the USA (number of installed stations)

antennas. In 1981, Equatorial developed a two-way VSAT network, based on its C-200 series, and offered the services to Farmers Insurance which became one of its main customers. Farmers Insurance equipped itself with over 2400 C-200 VSATs. Subsequently the company was bought by ASC Contel, which is now part of GTE Spacenet since 1990. Today the C-200 type VSAT has been renamed as Equastar.

1.7.2   VSAT development in the USA

In the USA, VSAT networks have been installed for transmission of voice, data, and video. Figure 1.25 shows the number of installed VSATs in the USA as a function of time. The initial growth rate appears to be about 50% per year. The expected growth rate from 1992 onwards is about 30% per year.

General Motors owns and operates the largest network with about 10 000 operating VSATs. Most networks have around 150 to 200 VSATs.

1.7.3   User categories in the USA

In the USA, VSAT main user categories are:

—automotive industry (General Motors, Chrysler, etc.)

—hotels (Holiday Inn, Days Inn, etc.)

—retail stores (Wal-mart, K-Mart, Penney, etc.)

—financial institutions (Prudential Bache Securities, etc.)

—news agencies (Reuters, etc.)

—religious institutions (LDS Church, etc.)

One also finds a number of small and medium size users with less than 100 terminals. About 95% of VSATs have transmit–receive capabilities. 5% are one-way VSATs used in data broadcast applications.

1.7.4 VSAT development in Europe

In Europe, the growth rate is presently (1994) of the order of 80% per year. But the number of installed VSATs is much smaller than in the USA as the result of a later development. Europe comprises many countries with different ruling bodies, and hence does not offer the large unified geographical area which has triggered favourably the use of VSAT technology in the USA. Most VSAT networks are national ones and the VSAT stations are most often owned and operated by the national telecommunications authority. The national authority ensures the maintenance of the network and rents the equipment to the customer. Only in a few cases are the VSATs owned by private companies and bought directly from the manufacturer by the user in agreement with the national telecom authority. Typical figures for Europe indicate about 30 hubs installed, with an average of 100 VSATs per hub, and an average network size of 40 VSATs [BUL93].

1.7.5 VSAT development in other countries

In India, two networks are currently operated:

—the National Informatics Center Network (NICNET) which is a national computer network with about 600 locations;

—the Remote Area Business Message Network (RABMIN) which is a public network operated by the national PTT. RABMIN uses the ARABSAT-1C satellite and supports services such as messaging, Email, telex and facsimile. It also provides gateways to packet switching networks.

In China, GTE Spacenet has constructed a VSAT network with about 2000 VSATs for the China National Petroleum and Chemical Co. of Beijing. The network costs about $10 million. It makes use of a Chinese satellite, and conveys data, voice and facsimile traffic, using a central hub located in Beijing. China is predicted to have 8000 two-way VSATs installed by 2002, more than any other country in the Asia/Pacific region.

In Russia, the news agency Itar-Tass relays news items from two 11 m earth stations located near Moscow over two INTELSAT satellites (AOR and IOR) to about 200 remote VSATs.

Other countries within the former Eastern Europe have developed a VSAT network as an alternative means to their poor terrestrial communications infrastructure. For instance, Poland has started in June 1993 an 80 VSAT network with a hub located near Warsaw. Hungary, Romania and the former Czechoslovakia have equipped themselves, under a European Space Agency (ESA) pilot project, with eight VSATs built by Alenia Spazio as part of a meshed network for village telephony, using a EUTELSAT satellite. Each VSAT has a capacity of 32 telephone lines.

VSAT activity is also growing rapidly in Latin and South America. More than 250 private satellite communications Ku-band networks have been set up in Mexico since 1987, using the Morelos satellites (launched in 1985) and now the Solidaridad satellite (launched in 1993). About 600 VSATs are in use in Argentina.

1.7.6   Regional VSAT market share

Figure 1.26 displays the present market share for two-way VSATs [BUL93]. North America dominates with about 80% of the market of ordered and installed VSATs, with Europe, Asia/Pacific and Latin America far behind with nearly equal 6% shares.

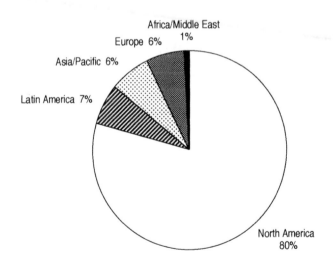

100% ≈ 120 000 VSATs ordered (out of which 60 000 are installed)

Figure 1.26   Regional market share for two-way VSATs as of 1993 [BUL93]

1.8   ECONOMIC ASPECTS

VSAT business faces competition in regions of the world where terrestrial networks are available. For the same grade of service, terrestrial digital data service networks and packet switched networks are probable contenders [CHA88]. Provided that desired network reliability, response time and throughput are achieved, then the economic comparison should be made on the basis of cost per month per site. Such comparisons with alternative solutions must be made on a case-by-case basis as they depend on many local factors, and also may be time dependent. In general terms a VSAT network is a cost effective solution if the unique property of the satellite to broadcast information is well used.

To illustrate the budget headings of a typical VSAT network and the impact of the number of VSATs, Tables 1.7 to 1.9 display the budget of three networks for a European scenario considering a typical interactive data application: Table 1.7 relates to a small network with 30 VSAT stations and a shared hub, Table 1.8 refers to a network of 300 VSAT stations and a dedicated mini-hub, and Table 1.9 relates to a network with 1000 VSAT stations and a dedicated large hub.

Table 1.7   Budget for a 30 VSAT network and a shared hub (European scenario)

	Cost per unit	Units (over 5 years)	Total	Cost/month
VSAT				
Equipment	10 000	30	300 000	5 000
Installation	1 000	30	30 000	500
Spare parts	1 000	30	30 000	500
Maintenance per VSAT per year	1 000	$30 \times 5 = 150$	150 000	2 500
Hub				
Lease cost per year	40 000	5	200 000	3333
Hub-to-host connection cost per year	20 000	5	100 000	1667
Satellite				
Bandwidth lease per year (1.25 MHz)	200 000	5	1 000 000	16 667
Licence				
One-time fee	8 000	1	8 000	133
Licence charge per VSAT per year	100	$5 \times 30 = 150$	15 000	250
Total cost			1 833 000	30 550
Cost per VSAT per month				**1018 ECU**

All costs in ECU (1 ECU ≈ $1.2).

Table 1.8   Budget for a 300 VSAT network and a dedicated mini-hub (European scenario)

	Cost per unit	Units (over 5 years)	Total	Cost per month
VSAT				
Equipment	8 000	300	2 400 000	40 000
Installation	1 000	300	300 000	5 000
Spare parts	1 000	300	300 000	5 000
Maintenance per VSAT per year	1 000	$300 \times 5 = 1\,500$	1 500 000	25 000
Hub				
Equipment and installation	100 000	1	100 000	1 667
Operation and maintenance per year	320 000	5	1 600 000	26 667
Satellite				
Bandwidth lease per year (2.5 MHz)	400 000	5	2 000 000	33 333
Licence				
One-time fee	8 000	1	8 000	133
Licence charge per VSAT per year	100	$5 \times 300 = 1\,500$	150 000	2 500
Total cost			8 358 000	139 300
Cost per VSAT per month				**464 ECU**

All costs in ECU (1 ECU ≈ $1.2).

Costs are given in ECU (European currency unit). The ECU is about $1.2. The cost per month per VSAT is calculated assuming a five-year amortisation. The 10 000 ECU equipment cost of a VSAT is applicable to a station equipped with a 1.8 m antenna, a 2 W transmitter power, and four output user ports. This cost figure corresponds to an order for a small number (30 units). A discount of 20% is applicable should 300 VSATs be ordered, and 30% for 1000 VSATs. This is reflected in Tables 1.8 and 1.9. Installation cost is taken as equal to one man-day plus professional expenses, or typically 1000 ECU per VSAT. Spare parts represent 10% of the equipment cost, and the annual maintenance cost is taken as equal to 1000 ECU per VSAT.

The VSAT transmits information at bit rate $R_b = 64\,kb/s$, with code rate $\rho = 1/2$. Therefore the transmitted bit rate is $R_c = 128\,kb/s$. Considering BPSK modulation, with spectral efficiency $\Gamma = 0.5\,b/s$ Hz, including guard bands, the carrier bandwidth is $B_{inb} = R_c/\Gamma = 250\,kHz$.

The access to the satellite is a time division multiple access (TDMA) and it is assumed that several VSAT stations share one inbound link.

Table 1.9 Budget for a 1000 VSAT network and a dedicated large hub (European scenario)

	Cost per unit	Units (over 5 years)	Total	Cost per month
VSAT				
Equipment	7 000	1 000	7 000 000	116 667
Installation	1 000	1 000	1 000 000	16 667
Spare parts	1 000	1 000	1 000 000	16 667
Maintenance per VSAT per year	1 000	$5 \times 1000 = 5\,000$	5 000 000	83 333
Hub				
Equipment and installation	1 000 000	1	1 000 000	16 667
Operation and maintenance per year	320 000	5	1 600 000	26 667
Satellite				
Bandwidth lease per year (3.5 MHz)	560 000	5	2 800 000	46 667
Licence				
One-time fee	8 000	1	8 000	133
Licence charge per VSAT per year	100	$5 \times 1000 = 5\,000$	500 000	8 333
Total cost			19 908 000	331 800
Cost per VSAT per month				**332 ECU**

All costs in ECU (1 ECU ≈ $1.2).

The outbound link is a time division multiplex (TDM) at information bit rate $R_b = 256$ kb/s, with code rate $= 1/2$. The transmitted bit rate is $R_c = 512$ kb/s and the utilised band is $B_{outb} = 1$ MHz.

Table 1.7 refers to a network with 30 VSATs, with one inbound link and one outbound link, so that the utilised transponder bandwidth is $B_1 = B_{inb} + B_{outb} = 1.25$ MHz.

Table 1.8 refers to a network with 300 VSATs, with six inbound links (50 VSATs per inbound link) and one outbound link, so that the utilised transponder bandwidth is $B_2 = 6B_{inb} + B_{outb} = 2.5$ MHz.

Table 1.9 refers to a network with 1000 VSATs, with ten inbound links (100 VSATs per inbound link) and one outbound link, so that the utilised transponder bandwidth is $B_3 = 10B_{inb} + B_{outb} = 3.5$ MHz.

A cost of 160 000 ECU per year is considered for the non-preemptible lease of 1 MHz of transponder bandwidth, which is typical of what is charged by EUTELSAT signatories in Europe. A US transponder is cheaper, and the impact of this tariff reduction will be examined.

Table 1.10 Sensitivity of cost per month per VSAT to space segment cost. The last column indicates the cost reduction factor under the assumption that transponder lease reduces from 160 000 ECUs per MHz per year to 46 000 ECUs per MHz per year

Type of network	Reference	Cost reduction factor
Shared hub, 30 VSATs	Table 1.7	0.61
Dedicated mini-hub, 300 VSATs	Table 1.8	0.83
Dedicated large hub, 1000 VSATs	Table 1.9	0.90

To operate and maintain a dedicated large hub or a mini-hub, a staff of eight people working round the clock in eight-hour shifts is considered with a cost per man-year of 40 000 ECU. Hence the annual staff cost in Tables 1.8 and 1.9 is 320 000 ECU. With a shared hub (Table 1.7) the staff is employed by the owner of the hub and the corresponding cost is charged to the client as part of the lease cost.

The licence structure is based on a one-time fee of 8000 ECU, reflecting the typical cost of the application to a national regulatory body within Europe, and a typical European annual fee per VSAT of 100 ECU. It can be noted that the one-time fee, although it may be considered as an expensive one, has little impact on the cost per month per VSAT.

The comparison of cost per month per VSAT from Tables 1.7 to 1.9 indicates the advantage gained from a large network. Unfortunately, in Europe, as of now, the scenario of Table 1.7 is the most likely one, as a result of the small demand and the difficulties encountered when planning to implement transborder networks. However, the comparison gives some insight into what the future would look like at a true European scale. From Table 1.9 one understands better the strong economic advantage achieved by North American VSAT networks.

The tables also indicate the influence of the space segment cost, which is higher for small size networks (see for instance Table 1.7). Therefore, it is of interest to examine the impact that a tariff reduction for satellite bandwidth would have on the cost per month per VSAT. Taking as a reference a typical lease tariff of space segment over the US of $2 million per year for a 36 MHz transponder, or equivalently 46 000 ECU per MHz per year, Table 1.10 indicates the cost reduction factor for the three above-mentioned networks. By mutiplying the cost-per-month value per VSAT as of Tables 1.7 to 1.9 by the corresponding cost reduction factor of the last column of Table 1.10, one obtains the cost per month per VSAT should the annual lease cost of 1 MHz of bandwidth reduce from the present European tariff to the present US value. This clearly shows how important an alignment with the US transponder lease tariffs would be for the European market of VSATs.

1.9   REGULATORY ASPECTS

Regulations in the field of VSAT networks entail several aspects, which will now be introduced and developed in more detail in Chapter 3, section 3.1

1.9.1 Standardisation

Standardisation relates to the activities of international bodies such as the International Standardisation Organisation (ISO), the International Electrotechnical Commission (IEC), and the International Telecommunication Union (ITU). The ITU comprises three sectors: the Radiocommunications Sector of the ITU (ITU-R, formerly CCIR), the Telecommunication Standardisation Sector (ITU-T) and the Development Sector (ITU-D). The Radiocommunications sector has issued five Recommendations dealing with VSAT radio frequency equipments [ITO93]:

—Rec 725: Technical characteristics for VSATs

—Rec 726: Maximum permissible level of spurious emissions from VSATs

—Rec 727: Cross polarisation isolation from VSATs

—Rec 728: Maximum permissible level of off-axis EIRP density from VSATs

—Rec 729: Control and monitoring function of VSATs

Within the same sector, recommendations regarding the interconnection between VSAT networks and public switched networks are being developed.

The main reasons for creating such recommendations are to provide a means of simplifying licensing procedures on a world-wide basis and to promote simple and economical use of VSAT networks.

In Europe, the European Telecommunications Standard Institute (ETSI) was founded in order to promote common European technical standards in order to initiate the conditions for an open telecommunications market in Europe [SAL93]. The ETSI involves a very large national participation which encompasses not only the European Community (EC) but also the European Free Trade Association (EFTA) countries and others, in increasing number, from the former Eastern Europe (Poland, Hungary, etc.). Members of ETSI are national administrations, public network operators, manufacturers, users and research bodies. They must be based in Europe, although non-European companies with European subsidiaries or offices may join. A Technical Committee (TC) dedicated to the creation of European Technical Standards (ETS) for Satellite Earth Stations (SES), named TC-SES, has issued a number of VSAT standards essentially in agreement with the ITU-R Recommendations:

—the ETS 300 157 provides specifications for the standardisation of the characteristics of receive-only VSATs operating as part of a network used for the distribution of data;

—the ETS 300 159 provides specifications for the standardisation of the characteristics of transmit/receive VSATs operating as part of a satellite network used for the distribution and/or exchange of data between users.

Both standards deal with protection of other users of the frequency spectrum, both satellite and terrestrial, from unacceptable interference. Also considered are requirements for safety (protection from insecure structures, electric shock, radio frequency radiation hazards, solar radiation focusing effects). Finally recommendations are given concerning protection of VSATs from interference produced by other radio systems.

Other VSAT standards issued by TC-SES are as follows:

—the ETS 300 160 is applicable to two-way VSATs operating in the framework of a satellite network for digital communication purposes as defined in the above ETS 300 159. In these networks there is a set of control and monitoring functions at each VSAT and a separate set of centralised control and monitoring functions;

—the ETS 300 160 deals with the control and monitoring functions at the VSAT while the requirements for the centralised control and monitoring functions are contained in the ETS 300 161. These functions aim at limiting interference to radio spectrum users due to fault conditions at VSATs that are only capable of transmitting baseband digital signals. It is required from these unattended VSATs that they be able:

(a) to monitor each of its processors involved in the manipulation of traffic and in control and monitoring functions,

(b) to receive properly the control information from the centralised control and monitoring facility and execute the corresponding commands,

(c) to suppress its transmission in case of misoperation in the transmit subsystem or in the event of a processor failure.

In the USA, the Federal Communications Commission (FCC) has based its 'blanket licensing' by reference to technical standards that aim at allowing the operation of a VSAT station in a two-degree orbit spacing situation [MIT93]. This has imposed special restrictions on VSAT transmission power, antenna size and side-lobe gain.

A similar approach has been used in Japan [FUJ93].

Further information on the background to standardisation is given in section 3.1. A detailed discussion of the development of US standards for VSAT networks is given in [MIT93].

Table 1.11 summarises the standards which prevail for the blanket licensing in the USA [MIT93] and Japan [FUJ93], and provides a comparison with those standards issued by ETSI [ITO92] [SAL93].

Table 1.11 Summary of existing regional standards [ITO92][FUJ93][SAL93]

Item	USA		Japan	ETSI
Frequency (GHz)	C-band	Ku-band up: 14–14.5* down; 11.7–12.2*	Ku-band up: 14–14.4* down: 12.5–12.75*	Ku-band up: 14–14.25 down: 12.5–12.75 10.7–11.7
Size	>4.1 m*	>1.2 m*	<2.4 m*	<3.7 m
Power	0.5 dBW/4kHz* analogue −2.7 dBW/4kHz* digital	−14 dBW/4kHz*	10 W*	—
Bandwidth (analogue)	<200 kHz*			<15 kHz*
Bit rate (digital)	<4.839 Mbit/s*	<512 kbit/s*	<3300 kbit/s*	<2.048 Mbit/s
Frequency tolerance	100 ppm		100 ppm*	Within nominated bandwidth
Spurious			100 μW at 10 W*	$+4 - 10 \log N^\dagger$ if carrier on −21 if carrier off dBW/100 kHz
Inland emission			at least 20 dB below maximum carrier power	
Outband emission			at least 20 dB below maximum carrier power or <100 μW	
Modulation			FM or PM* including PSK	
Cross pol			>27 dB	>27 dB
Off-axis	$15 - 25 \log \theta^*$ dBW/4KHz		$33 - 25 \log \theta$	$33 - 25 \log \theta - 10 \log N^\dagger$
EIRP	(assume −14 dBW/4kHz input and $29 - 25 \log \theta$ for an antenna)		$2.5° \leqslant \theta < 7°$ 12 $7° \leqslant \theta < 9.2°$ $36 - 25 \log \theta$ $9.2° \leqslant \theta < 48°$ −6 $48° \leqslant \theta < 180°$ dBW/40kHz	$2.5° \leqslant \theta < 7°$ $12 - 10 \log N$ $7° \leqslant \theta < 9.2°$ $36 - 25 \log \theta - 10 \log N$ $9.2° \leqslant \theta < 48°$ $-6 - 10 \log N$ $48° \leqslant \theta < 180°$ dBW/40kHz

(*continued*)

Item	USA	Japan	ETSI
Antenna pattern (dBi)	$29 - 25 \log \theta$		$29 - 25 \log \theta$ $2.5° \leqslant \theta < 7°$ 8 $7° \leqslant \theta \leqslant 9.2°$ $32 - 25 \log \theta$ $9.2° \leqslant \theta < 48°$ -10 $48° \leqslant \theta < 180°$
Typical signal			Digital
Protocol interface			X25
Control/ monitor		Tx: cease when malfunction/ Central control Enable Disable	Central control Enable Disable/ status monitoring

*For 'blanket licensing': simplified licensing procedure based on type approval of a station.
†N is the maximum number of VSATs which are expected to transmit simultaneously in the same carrier frequency band. This number shall be indicated by the manufacturer.

1.9.2   Licensing of operation

Licensing of operation is delivered by the national telecommunications authority where any earth station as a part of a network, be it the hub, a control station or a VSAT, is planned to be installed and operated. A licence usually entails the payment of a licence fee, which is most often of two types: a one-time fee for the licensing work and an annual charge per station.

The matter is simpler when the network is national. For transborder networks, licences must be obtained from the different national authorities where the relevant earth stations are planned to be installed and operated, and rules differ from one country to another.

1.9.3   Access to the space segment

This deals with the performance and operational procedures that satellite operators request from earth station operators for transmission of carriers to, and from, their satellite transponders. Most often such requirements are based on ITU-R Recommendations, but some satellite operators may impose special constraints.

1.9.4   Local regulations

Finally, one should also mention the compliance of various dish aspects, e.g. size, colour, shape, with local regulations dealing with environmental protection, safety and landlord permission to install roof mounted antennas.

1.10   CONCLUSIONS

The conclusion to this introduction is an appropriate opportunity to summarise the advantages and drawbacks of VSAT networks.

1.10.1   Advantages

1.10.1.1   *Point-to-multipoint and point-to-point communications*

A VSAT network offers communications between remote terminals. As a result of the power limitation resulting from the imposed small size and low cost of the remote station, a VSAT network is most often star shaped with remotes linked to a larger station called the hub. This star configuration reflects well the structure of information flow within most large organisations which have a point of central control where the hub can be installed. The star configuration itself is not a severe limitation to the effectiveness of a VSAT network as point-to-point communications, which would conveniently be supported by a meshed network, can still be achieved via a double hop, using the hub as a central switch to the network.

1.10.1.2   *Asymmetry of data transfer*

As a result of its asymmetric configuration, a star shaped network displays different capacities on the inbound link and on the outbound link. This may be an advantage also, considering the customer's need for asymmetric capacities in most of his applications. Should he use leased terrestrial lines which are inherently symmetric, i.e. offering equal capacity in both directions, the customer would have to pay for unused capacity.

1.10.1.3   *Flexibility*

A VSAT network inherently provides quick response time for network additions and reconfigurations (one or two days) as a result of the easy displacement and installation of a remote station.

1.10.1.4   Private corporate networks

A VSAT network offers its operator end-to-end control over transmission quality
and reliability. It also protects him from possible and unexpected tariff fluctuations, by offering price stability, and the possibility to forecast its communication
expenses. Therefore it is an adequate support to private corporate networks.

1.10.1.5   Low bit error rate

Bit error rate usually encountered on VSAT links is typically 10^{-7}.

1.10.1.6   Distance insensitive cost

The cost of a link in a VSAT network is not sensitive to distance. Hence cost
savings are expected if the network displays a large number of sites and a high
geographical dispersion.

1.10.2   Drawbacks

1.10.2.1   Interference sensitive

A radio frequency link in a VSAT network is subject to interference as a result of
the small earth station antenna size.

1.10.2.2   Eavesdropping

As a result of the large coverage of a geostationary satellite, it may be easy for
eavesdroppers to receive a downlink carrier and access the information content
by demodulating the carrier. Therefore to prevent unauthorised use of the
information conveyed on the carrier, encryption may be mandatory. A proposal
for a cryptosystem for VSAT networks is discussed in [SON92].

1.10.2.3   Loss of transponder may lead to loss of network

The satellite is a single point failure. Should the transponder which relays the
carrier fail, then the complete VSAT network is out of order. Communication links
can be restored by using a spare transponder. With a spare colocated on the same
satellite, a mere change in frequency or polarisation puts the network back to
operation. However, should this transponder be located on another satellite, this

may impose intervention on each site to repoint the antenna, and this may take some time.

1.10.2.4 Propagation delay; (double hop = 0.5 s)

The propagation time from remote to remote in a star shaped network imposes a double hop with its associated delay of about half a second. This may prevent the use of voice communications, at least with commercial standards.

2 USE OF SATELLITES FOR VSAT NETWORKS

It is not so important for someone who is interested in VSAT networks to know a lot about satellites. However, a number of factors relative to satellite orbiting and satellite–earth geometry influence the operation and performance of VSAT networks. For instance, the relative position of the satellite with respect to the VSAT at a given instant determines the orientation of the VSAT antenna and also the carrier propagation delay value. The relative velocity of the satellite with respect to the earth station receiving equipment induces Doppler shifts on the carrier frequency that must be tracked and compensated for. This impacts on the specifications and the design of earth station receivers. For a geostationary satellite, which is supposed to be in a fixed position relative to the Earth, one may believe that once the antenna has been properly pointed towards that position at the time of its installation, the adequate orientation is established once and for all. Actually, as a result of satellite orbital perturbations, there is no such thing as a geostationary satellite, and residual motions induce antenna depointing and hence antenna gain losses which affect the link performance.

Therefore it is worth mentioning these aspects, and this is the aim of this chapter. Orbit definition and parameters will be presented in the general case, with the ulterior motive to give the reader some conceptual tools that would be handy should VSAT networks be used some day in conjunction with non-geostationary satellite systems. However, as current VSAT networks use geostationary satellites, the bulk of the chapter will consider this specific scenario. Many of the considerations developed in this chapter will be used in the following ones.

Before orbital aspects are dealt with, the role of the satellite and some related topics will first be introduced as an encouragement to the reader.

2.1 INTRODUCTION

2.1.1 The relay function

Satellites relay the carriers transmitted by earth stations on the ground to other earth stations, as illustrated in Figure 2.1. Therefore, satellites act similarly to

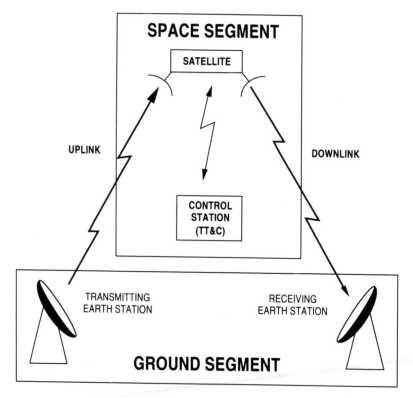

Figure 2.1 Architecture of a satellite system

microwave terrestrial relays installed on the top of hills or mountains to facilitate long distance radio frequency links. Here the satellite, being at a much higher altitude than any terrestrial relay, is able to link distant earth stations, even from continent to continent.

Figure 2.1 indicates that the earth stations are part of what is called the *ground segment*, while the satellite is part of the *space segment*. The space segment also comprises all the means to operate the satellite, as for instance the stations which monitor the satellite status by means of telemetry links, and control it by means of command links. Such links are sometimes called TTC (Telemetry, Tracking and Command) links.

The satellite roughly consists of a platform and a payload. The platform consists of all subsystems that allow the payload to function properly, namely:

—the mechanical structure which supports all equipments in the satellite;

—the electric power supply, consisting of the solar panels and the batteries used as supply during eclipses of the sun by the Earth and the Moon;

—the attitude and orbit control, with sensors and actuators;

—the propulsion subsystem;

—the onboard TTC equipment.

The payload comprises the satellite antennas and the electronic equipment for amplifying the uplink carriers. These carriers are also frequency converted to the frequency of the downlink. Frequency conversion avoids unacceptable interference between uplinks and downlinks.

Figure 2.2 shows the general architecture of the payload. The receiver (RX) encompasses a wide band amplifier and a frequency downconverter. The input multiplexer (IMUX) splits the incoming carriers into groups within several sub-bands, each group being amplified to the power level required for transmission by a high power amplifier, generally a travelling wave tube (TWT). The different groups of carriers are then combined in the output multiplexer (OMUX) and forwarded to the transmitting antenna. The channels associated with the sub-bands of the payload from IMUX to OMUX are called *transponders*. The advantage of splitting the satellite band is three-fold:

—each transponder TWT amplifies a reduced set of carriers, hence each carrier benefits from a larger share of the limited amount of power available at the output of the TWT;

—the transponder TWT operates in a non-linear mode when driven near saturation. Saturation is desirable because the TWT then delivers more power to the amplified carriers than when operated in a backed-off mode, away from saturation. However, amplifying multiple carriers in a non-linear mode generates intermodulation, which acts as transmitted noise on the downlink. Less intermodulation noise power is transmitted with a reduced set of amplified carriers within each TWT;

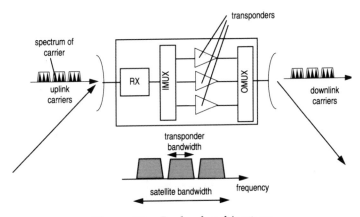

Figure 2.2 Payload architecture

—reliability is increased, as the failure of one TWT does not imply an overall satellite failure and each TWT can be backed up.

Typical values of bandwidth for a transponder are 36 MHz, 45 MHz, and 72 MHz. However, there is no established standard. The TWT power is typically a few tens of watts. Some satellites are now equipped with solid state power amplifiers (SSPA) instead of TWTs.

Figure 2.2 does not indicate any back-up equipment. To actually ensure the required reliability at the end of life of the satellite, some redundancy is built into the payload: for instance, the receiver is usually backed up with a redundant unit, which can be switched on in case of failure of the allocated receiver. The transponders are also backed up by a number of redundant units: a popular scheme is the ring redundancy, where each IMUX output can be connected to any of several transponders, with a similar arrangement between the transponder outputs and the OMUX inputs.

2.1.2 Transparent and regenerative payload

A satellite payload is transparent when the carrier is amplified and frequency downconverted without being demodulated. The frequency conversion is then performed by means of a mixer and a local oscillator as indicated in Figure 2.3: the carrier at a frequency equal to the uplink frequency f_U minus the local oscillator frequency f_{LO} is usually selected by filtering at the output of the mixer, and the local oscillator frequency is tuned so that the resulting frequency corresponds to the desired downlink frequency f_D. For instance, an uplink carrier at frequency $f_U = 14.25\,\text{GHz}$ mixed with a local oscillator frequency $f_{LO} = 1.55\,\text{GHz}$ results in a downlink carrier frequency $f_D = 12.7\,\text{GHz}$.

A transparent payload makes no distinction between uplink carrier and uplink noise, and both signals are forwarded on the downlink. Therefore, at the earth station receiver, one gets the downlink noise together with the uplink retransmitted noise.

A regenerative payload entails on-board demodulation of the uplink carriers. On-board regeneration is most conveniently performed on digital carriers. The bit stream obtained from demodulation of a given uplink carrier is then used to modulate a new carrier at downlink frequency. This carrier is noise-free, hence

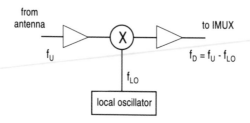

Figure 2.3   Receiver for a transparent satellite

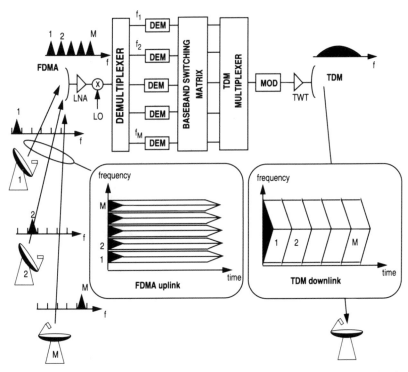

Figure 2.4 Regenerative satellite payload with multiplexed transmission on the downlink

a regenerative payload does not retransmit the uplink noise on the downlink. The overall link quality is therefore improved. Moreover, intermodulation noise can be avoided as the satellite channel amplifier is no longer requested to operate in a multicarrier mode. Indeed, several bit streams at the output of various demodulators can be combined into a time division multiplex (TDM) which modulates a single high rate downlink carrier. This carrier is amplified by the channel amplifier which can be operated at saturation without generating intermodulation noise as the carrier it amplifies is unique. This concept is illustrated in Figure 2.4.

It should be emphasised that today's commercial satellites are not equipped with regenerative payloads but only with transparent ones. Only a few experimental satellites such as NASA's Advanced Communications Technology Satellite (ACTS) and the Italian ITALSAT incorporate a regenerative payload. The chances that regenerative payloads will be used in the future to support VSAT networking for commercial services is discussed in Chapter 6, section 6.3.

2.1.3 Coverage

The coverage of a satellite payload is determined by the radiation pattern of its antennas. The receiving antenna and the transmitting antenna may have different patterns and hence there may be a different coverage for the uplink and the

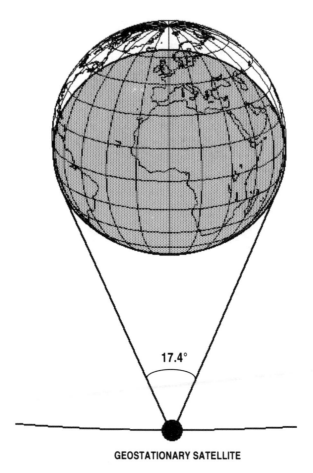

GEOSTATIONARY SATELLITE

Figure 2.5 Global coverage

downlink. The coverage is usually defined by a specified minimum value of the antenna gain: for instance, the 3 dB coverage corresponds to the area defined by a contour of constant gain value 3 dB lower than the maximum gain value at antenna boresight. This contour defines the *edge of coverage*.

There are four types of coverage:

—*Global coverage*:   the pattern of the antenna illuminates the largest possible portion of the surface of the Earth as viewed from the satellite (Figure 2.5). A geostationary satellite sees the earth with an angle equal to 17.4°. Selecting the beamwidth of the antenna as 17.4° imposes that the maximum gain at boresight is 20 dBi, and then the gain at edge of the minus 3 dB coverage is 17 dBi.

—*Zone coverage*:   an area smaller than the global coverage area is illuminated (Figure 2.6). The coverage area may have a simple shape (circle or ellipse) or a more complex shape (contoured beam). For a typical zone coverage the antenna beamwidth is of the order of 5°. This imposes a maximum gain at boresight of 30 dBi, and a gain at edge of the minus 3 dB coverage of 27 dBi.

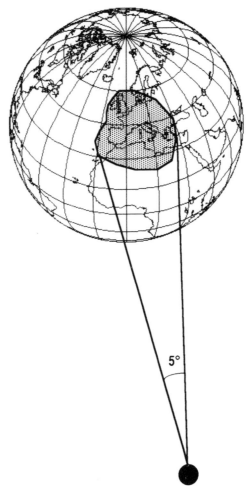

GEOSTATIONARY SATELLITE

Figure 2.6 Zone coverage

—*Spot beam coverage*: an area much smaller than the global coverage area is illuminated. The antenna beamwidth is of the order of 2° (Figure 2.7). Considering a 1.7° beamwidth imposes a maximum gain at boresight of 40 dBi and a gain at edge of the minus 3 dB coverage of 37 dBi.

—*Multibeam coverage*: a spot beam coverage has the advantage of higher antenna gain than any other type of coverage previously discussed, but it can service only the limited zone within its coverage area. A service zone larger than the coverage area of a spot beam can still be serviced with high antenna gain thanks to a multibeam coverage made of several individual spot beams. An example of such a coverage with adjacent spot beams is shown in Figure 2.8. This requires a multibeam satellite payload with more complex antenna farms. Maintaining interconnectivity between all stations of the service zone also

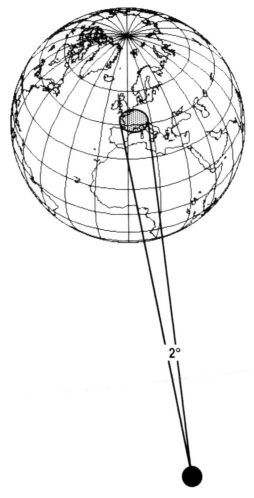

GEOSTATIONARY SATELLITE

Figure 2.7   Spot beam coverage

implies a more complex payload architecture than that considered in Figure 2.2. Interconnectivity between stations implies that beams be interconnected: this can be achieved either by permanent connections from the uplink beams to the downlink ones, as illustrated in Figure 2.9, or by temporary connections established through an on-board switching matrix, as shown in Figure 2.10.

Permanent connections entail a larger number of transponders than on-board switching. On-board satellite switching requires that earth stations transmit bursts of carriers, synchronous to the satellite switch state sequence, in such a way that they arrive at the satellite exactly when the proper uplink beam to downlink beam connection is established. More details on the operation of such multibeam satellite systems can be found in [MAR93, Chapter 5].

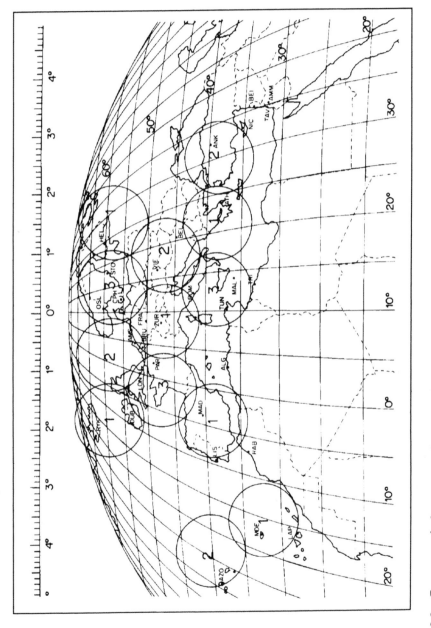

Figure 2.8 Coverage of a larger zone than covered by a single beam using a multibeam satellite. (Reproduced from [MAR93] by permission of John Wiley & Sons Ltd)

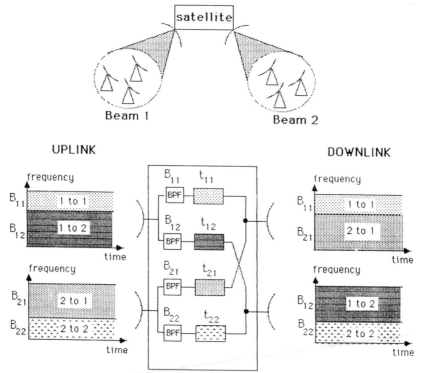

Figure 2.9 Interconnectivity of beams by permanent connections. (Reproduced from [MAR93] by permission of John Wiley & Sons Ltd)

Usually the extension of a VSAT network is small enough for all VSATs and the hub station to be located within one beam.

2.1.4   Impact of coverage on satellite relay performance

The relay function of the satellite as described in section 2.1.1 entails adequate reception of uplink carriers and transmission of downlink carriers. As will be demonstrated in Chapter 5, the ability of the satellite payload to receive uplink carriers is measured by the figure of merit G/T of the satellite receiver, and its ability to transmit is measured by its Effective Isotropic Radiated Power (EIRP). Those characteristics are defined in more detail in Chapter 5. Basically, G/T is the ratio of the receiving satellite antenna gain to the uplink system noise temperature, and the EIRP is the product of the transmitting satellite antenna gain G_T and the power P_T fed to the antenna by the transponder amplifier. Therefore, both parameters are proportional to the satellite antenna gain.

The specified values of G/T and EIRP are to be considered at edge of coverage. Usually the edge of coverage is defined by the contour on the Earth corresponding to a constant satellite antenna gain, say 3 dB below the gain G_{max} at boresight.

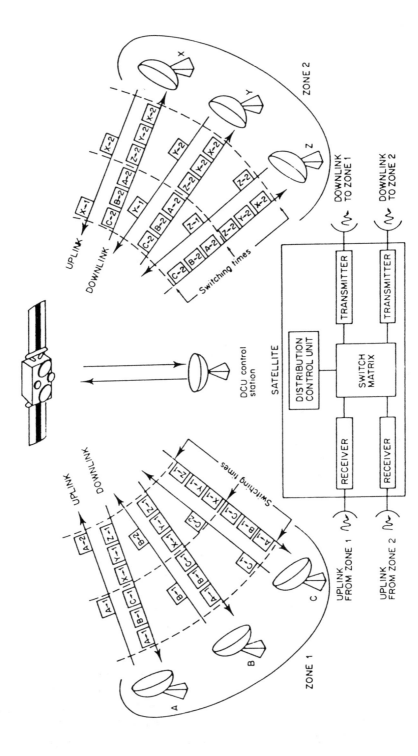

Figure 2.10 Interconnectivity of beams by temporary connections. (Reproduced from [MAR93] by permission of John Wiley & Sons Ltd)

Now the maximum satellite antenna gain, G_{max}, as obtained at boresight, is inversely proportional to the square of its half-power beamwidth θ_{3dB}:

$$G_{max} = \frac{29\,000}{\theta_{3dB}^2}$$

or

$$G_{max}(dBi) = 44.6 - 20\log\theta_{3dB} \tag{2.1}$$

Hence, one can consider that the specified values of G/T and EIRP are conditioned by the value of the satellite antenna gain at edge of coverage G_{eoc}, given by:

$$G_{eoc} = \frac{G_{max}}{2}$$

or

$$G_{eoc}(dBi) = G_{max}(dBi) - 3\ dB \tag{2.2}$$

From (2.2) and (2.1), it can be seen that the specified values of G/T and EIRP at edge of coverage are conditioned by the satellite antenna beamwidth θ_{3dB}: the larger the beamwidth, the lower the G/T and EIRP.

So, the coverage of the satellite influences its relaying performance in terms of G/T and EIRP. A global coverage leads to smaller values of satellite G/T and EIRP, compared to a spot beam coverage. Should the VSAT network be included in a single satellite beam, then the larger its geographical dispersion, the poorer the satellite performance: this has to be compensated for by installing larger VSATs. For networks comprised of highly dispersed VSATs, say spread over several continents, the advantages of simple networking in terms of easy interconnectivity by placing all VSATs within a single beam have to be weighed against the cost of increasing the size of the VSATs, which might not be necessary by accepting to service the network with a multibeam satellite, at the expense, however, of a more complex network operation.

2.1.5 Frequency reuse

Frequency reuse consists of using the same frequency band several times in such a way as to increase the total capacity of the network without increasing the allocated bandwidth.

Frequency reuse can be achieved within a given beam by using polarisation diversity: two carriers at same frequency but with orthogonal polarisations can be discriminated by the receiving antenna according to their respective polarisation. With multibeam satellites the isolation resulting from antenna directivity can be exploited to reuse the same frequency band in different beams.

Figure 2.11 compares the principle of frequency reuse (a) by orthogonal polarisation, and (b) by angular beam separation. In both cases the bandwidth allocated to the system is B. The system uses this bandwidth B centred on frequency f_U for the uplink and on the frequency f_D for the downlink. In the case of

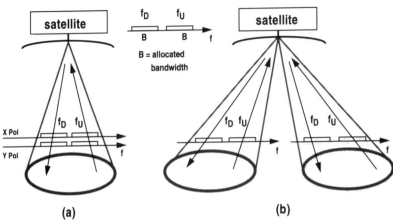

Figure 2.11 Frequency reuse; (a) by orthogonal polarisation; (b) by angular separation of the beams in a multibeam satellite system

frequency reuse by orthogonal polarisation, the bandwidth B can only be reused twice. In the case of reuse by angular separation, the bandwidth B can be reused for as many beams as the permissible beam to beam interference level allows. Both types of frequency reuse can be combined.

2.2 ORBIT

2.2.1 Newton's universal law of attraction

Satellites orbit the earth in accordance with Newton's universal law of gravitation: two bodies of mass m and M attract each other with a force which is proportional to their masses and inversely proportional to the square of the distance, r, between them:

$$F = GM\frac{m}{r^2} \quad (N) \tag{2.3}$$

where G (gravitational constant) $= 6.672 \times 10^{-11}\,\mathrm{m^3/kg\,s^2}$.

As the mass of the Earth is $M_e = 5.974 \times 10^{24}\,\mathrm{kg}$, the product GM_e for an earth orbiting body has a value

$$\mu = GM_e = 3.986 \times 10^{14}\,\mathrm{m^3/s^2}$$

From Newton's law, the following results can be derived, which actually were formulated prior to Newton's works by Kepler from his observation of the movement of the planets around the sun:

—the trajectory of the satellite in space, called its orbit, lies in a plane containing the centre of the Earth: for communication satellites, the orbit is selected to be an ellipse and one focus is the centre of the Earth. Should the orbit be circular, then the orbit centre coincides with the Earth's centre;

—the vector from the centre of the Earth to the satellite sweeps equal areas in equal times;

—the period T of revolution of the satellite around the Earth is given by:

$$T = 2\pi \sqrt{\frac{a^3}{\mu}} \quad \text{(seconds)} \tag{2.4}$$

where a is the semi-major axis of the ellipse (in meters).

2.2.2 Orbital parameters

Six parameters are required to determine the position of the satellite in space (Figure 2.12: [MAR93, Figure 7.4, p. 231]):

—two parameters for the determination of the plane of the orbit: the inclination of the plane (i) and the orbit right ascension of the ascending node (Ω);

—one parameter for positioning the orbit in its plane: the argument of the perigee (ω);

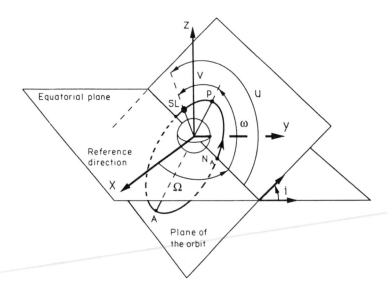

Figure 2.12 Positioning of satellite in space. (Reproduced from [MAR93] by permission of John Wiley & Sons Ltd)

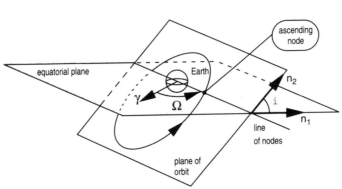

Figure 2.13 Orbit plane positioning: Ω, i

—two parameters for the shape of the orbit: the semi-major axis (a) of the ellipse, and its eccentricity (e);

—one parameter for the positioning of the satellite on the elliptic curve: the true anomaly (v).

2.2.2.1 Plane of the orbit (Figure 2.13)

The plane of the orbit is obtained by rotating the Earth's equatorial plane about the *line of nodes* of the orbit. The nodes are the intersections of the orbit with the equatorial plane of the Earth. There is one ascending node where the satellite crosses the equatorial plane from south to north, and one descending node where the satellite crosses the equatorial plane from north to south. The rotation angle about the line of nodes is i, defined as the *inclination of the orbital plane*. This angle is counted positively in the forward direction between $0°$ and $180°$ between the normal n_1 (directed towards the east) to the line of nodes in the equatorial plane, and the normal n_2 (in the direction of the satellite velocity) to the line of nodes in the orbital plane.

The line of nodes must be referenced to some fixed direction in the equatorial plane. The commonly used reference direction is the line of intersection of the Earth's equatorial plane with the plane of the ecliptic, which is the orbital plane of the Earth around the sun (Figure 2.14). This line maintains a fixed direction in space with time, called the *direction of the vernal point* γ. Actually, as a result of some irregularities in the rotation of Earth, with its axis experiencing nutation, the direction of the vernal point is not perfectly fixed with time. Therefore the reference direction is taken as the direction of the vernal point at some instant, usually noon on January 1, year 2000, designated as γ_{2000}. The angle which defines the direction of the line of nodes is the *right ascension of the ascending node* Ω: it is counted positively from $0°$ to $360°$ in the forward direction in the equatorial plane about the Earth's axis.

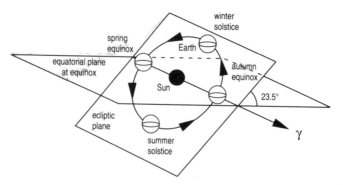

Figure 2.14 The direction of the vernal point γ is used as the reference direction in space

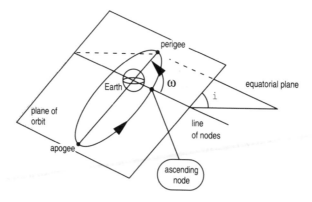

Figure 2.15 Positioning the orbit in its plane: the argument of the perigee (ω)

2.2.2.2 *Positioning the orbit in its plane (Figure 2.15)*

The centre of the Earth is one of the focuses of the elliptical orbit. Therefore, the major axis of the ellipse passes through the centre of the Earth. The direction of the perigee in the plane of the orbit is determined by the *argument of the perigee* ω, which is the angle, with vertex at the centre of the Earth, taken positively from 0° to 360° in the direction of the motion of the satellite between the direction of the ascending node and the direction of the perigee. The *perigee* is the point of the orbit that is nearest to the centre of the Earth. At the opposite point of the major axis is the *apogee*, which is the point of the orbit that is farthest from the centre of the Earth.

2.2.2.3 *Shape of the orbit (Figure 2.16)*

The shape of the orbit is determined by its *eccentricity, e,* and the length, *a,* of its *semi-major axis*. The eccentricity is given by:

$$e = \frac{c}{a}$$

(2.5)

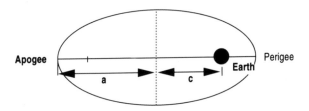

Figure 2.16   Defining the shape of the orbit: $a, e = c/a$

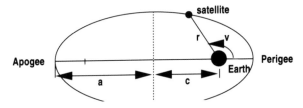

Figure 2.17   Positioning the satellite in its orbit

where c is the distance from the centre of the ellipse to the centre of the Earth. For a circular orbit the eccentricity is zero, and the centre of the Earth is the centre of the circular orbit.

The distance from the centre of the Earth to the apogee is $a(1 + e)$, and the distance from the centre of the Earth to the perigee is $a(1 - e)$.

2.2.2.4   *Positioning the satellite in its orbit (Figure 2.17)*

The position of the satellite in its orbit is conveniently defined by the *true anomaly*, v, which is the angle with vertex at the centre of the Earth counted positively in the direction of movement of the satellite from $0°$ to $360°$, between the direction of the perigee and the direction of the satellite.

The distance from the centre of the Earth to the satellite is given by:

$$r = a \frac{1 - e^2}{1 + e \cos v} \quad (\text{m}) \tag{2.6}$$

The satellite velocity is given by:

$$V = \mu^{1/2} \left(\frac{2}{r} - \frac{1}{a} \right)^{1/2} \quad (\text{m/s}) \tag{2.7}$$

2.3   THE GEOSTATIONARY SATELLITE

2.3.1   Orbit parameters

A geostationary satellite proceeds in a circular orbit ($e = 0$) in the equatorial plane ($i = 0°$). The angular velocity of the satellite is the same as that of the Earth, and in

Table 2.1   Characteristics of a geostationary satellite orbit

Eccentricity (e)	0
Inclination of orbit plane (i)	$0°$
Period (T)	23 h 56 min 4 s = 86 154 s
Semi-major axis (a)	42 164 km
Satellite altitude (R_o)	35 786 km
Satellite velocity (V_s)	3075 m/s

the same direction (direct orbit), as illustrated in Figure 1.4. To a terrestrial observer, the satellite seems to be fixed in the sky.

The above conditions impose the period of the circular orbit, T, to be equal to the duration of a sidereal day, that is the time it takes for the Earth to rotate $360°$. Hence $T = 23$ h 56 min 4 s = 86 164 s. From expression (2.4) one can calculate the semi-major axis, a, of the orbit which identifies the radius of the orbit. One obtains $a = 42\,164$ km. Subtracting from this value the Earth radius $R_e = 6378$ km, one obtains the satellite altitude $R_o = a - R_e = 35\,786$ km. The satellite velocity V_s can be calculated from expression (2.7) selecting $r = a$. It results in $V_s = 3075$ m/s.

Table 2.1 summarises the characteristics of a geostationary satellite orbit.

2.3.2   Launching the satellite

The principle of launching a satellite into orbit is to provide it with the appropriate velocity at a specific point of its trajectory in the plane of the orbit, starting from the launching base on the Earth surface. This usually requires a launch vehicle for the take-off, and an on-board specific propulsion system.

With a geostationary satellite, the orbit aimed at is circular, in the equatorial plane, and it is attained by an intermediate orbit called the *transfer orbit*. This is an elliptic orbit with perigee at an altitude of about 200 km, and apogee at the altitude of the geostationary orbit (35 786 km). Most conventional launch vehicles (Ariane, Delta, Atlas Centaur) inject the satellite into the transfer orbit at its perigee, as shown in Figure 2.18.

At this point, the launch vehicle must communicate a velocity $V_p = 10\,234$ m/s to the satellite (for a perigee at 200 km). Then the satellite is left to itself and proceeds forward in the transfer orbit. When arriving at the apogee of the transfer orbit, the satellite propulsion system is activated and a velocity impulse is given to the satellite. This increases its velocity to the required velocity for a geostationary orbit, that is $V_s = 3075$ m/s. The satellite orbit now is circular, and the satellite has the proper altitude.

Note the advantage of a launch towards the east as the launch vehicle benefits from the velocity introduced into the trajectory by the rotation of the Earth.

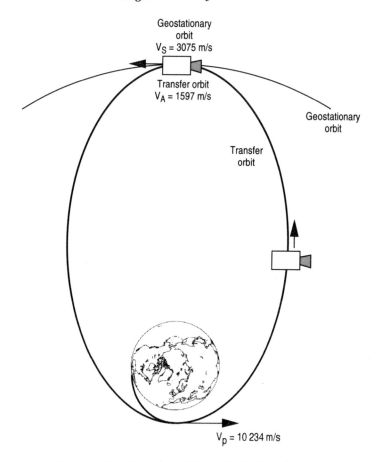

Geostationary
orbit
V_S = 3075 m/s

Transfer orbit
V_A = 1597 m/s

Geostationary
orbit

Transfer
orbit

V_p = 10 234 m/s

Figure 2.18 Transfer orbit and injection phases

In practice, there are some slight deviations to the above procedure:

—The launch base may not be in the equatorial plane. The launch vehicle follows a trajectory in a plane which contains the centre of the Earth and the launch base (Figure 2.19). The inclination of the orbit is thus greater than or equal to the latitude of the launching base, unless the trajectory is made non-planar, but this would induce mechanical constraints and an additional expense of energy. So the normal procedure is to have it planar. Should the launch base not be on the equator, then the transfer orbit and the final geostationary satellite orbit are not in the same plane, and an *inclination correction* has to be performed. This correction requires a velocity increment to be applied as the satellite passes through one of the nodes of the orbit such that the resultant velocity vector, V_s, is in the plane of the equator, as indicated in Figure 2.20. For a given inclination correction, the velocity impulse ΔV to be applied increases with the velocity V_A of the satellite. The correction is thus performed at the apogee of the transfer orbit where V_A is minimum, at the same time as circularisation.

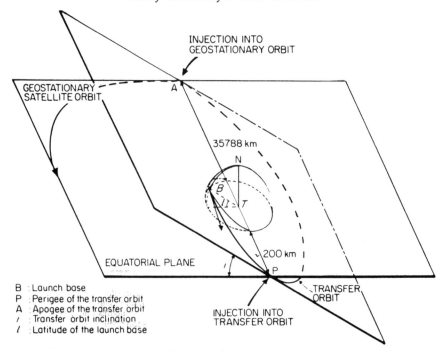

Figure 2.19 Sequence for launch and injection into transfer and geostationary orbit when the launch base in not in the equatorial plane. [(Reproduced from [MAR93] by permission of John Wiley & Sons Ltd)

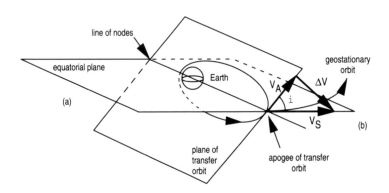

Figure 2.20 Inclination correction: (a) transfer orbit plane and equatorial plane; (b) required velocity increment (value and orientation) in a plane perpendicular to the line of nodes

—A precise determination of the transfer orbit parameters requires *trajectory tracking* during several orbits. Hence, the satellite propulsion system is activated only after several transfer orbit periods.

—The injection into geostationary orbit does not necessarily take place in the meridian plane of the Earth where the geostationary satellite is to be positioned for operation. To reach this position, a relative non-zero small angular velocity between the satellite and the Earth must be kept so that the satellite undergoes a longitudinal drift. This leads to injecting the satellite from transfer orbit into a circular orbit, called *drift orbit*, with a slightly different altitude than that of the geostationary satellite orbit. Once the satellite has reached the intended station longitude, a correction is initiated by activating the thrusters of the satellite orbit control system.

2.3.3 Distance to the satellite

The distance from an earth station to the satellite impacts on the propagation time of the radio frequency carrier and hence on the delay for information delivery (see Chapter 4, section 4.6). It also conditions the path loss which intervenes in the link budget calculation (see Chapter 5).

Figure 2.21 displays the geometry of the position of the earth station with respect to the satellite.

If we denote by l the geographical latitude of the earth station, and L the difference in longitude between that of the earth station and that of the satellite meridian, the distance R from the satellite to the earth station is then given by:

$$R = \sqrt{R_e^2 + (R_o + R_e)^2 - 2R_e(R_o + R_e)\cos \Phi} \quad \text{(m)} \tag{2.8}$$

where:

R_e = Earth radius = 6378 km
R_o = satellite altitude = 35 786 km
$\cos \Phi = \cos l \cos L$

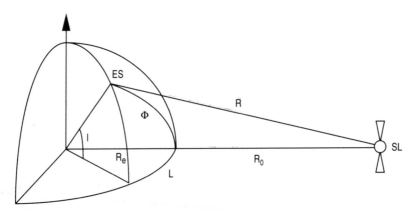

Figure 2.21   Relative position of the earth station (ES) with respect to the satellite (SL)

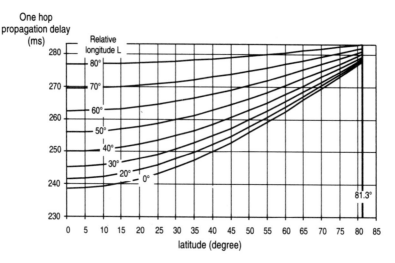

Figure 2.22 Single hop propagation delay as a function of the earth station latitude, *l*, and its relative longitude, *L*, with respect to the geostationary satellite meridian

With the above numerical values, equation (2.8) can be written as:

$$R = R_o\sqrt{1 + 0.42(1 - \cos \Phi)} \quad \text{(m)} \tag{2.9}$$

2.3.4 Propagation delay

The single hop propagation delay (from earth station to earth station) is given by:

$$T_p = 2\frac{R}{c} = 2\frac{R_o}{c}\sqrt{1 + 0.42(1 - \cos \Phi)} \quad \text{(s)} \tag{2.10}$$

where *c* is the velocity of light $= 3 \times 10^8$ m/s.
   Figure 2.22 displays T_p as a function of *l* and *L*.

2.3.5 Azimuth and elevation angles

In order to point an earth station antenna towards a geostationary satellite, one needs to know the azimuth (*Az*) and the elevation (*E*) angles. These angles are defined as follows (Figure 2.23):

—The azimuth angle *Az* is the rotation angle about a vertical axis through the earth station counted clockwise from the geographical north which brings the antenna boresight into the vertical plane which contains the satellite. This plane contains the centre of the Earth, the earth station and the satellite. The value of *Az* is obtained by means of an intermediate parameter, *a*, determined from the

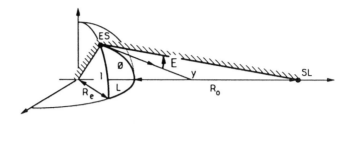

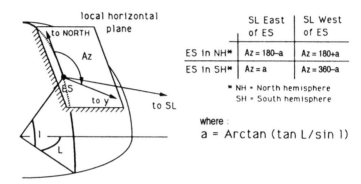

	SL East of ES	SL West of ES
ES in NH*	Az = 180−a	Az = 180+a
ES in SH*	Az = a	Az = 360−a

* NH = North hemisphere
SH = South hemisphere

where :
a = Arctan (tan L/sin l)

Figure 2.23 Definition of azimuth and elevation angles (ES: earth station, SL: satellite) (Reproduced from [MAR93] by permission of John Wiley & Sons Ltd)

family of curves of Figure 2.24 and used to calculate Az according to the table inserted in the figure [SMI72]. The curves are obtained from the following expression which can be used for greater accuracy:

$$a = \arctan\left(\frac{\tan L}{\sin l}\right) \quad \text{(degrees)} \tag{2.11}$$

—The elevation angle E is the rotation angle about a horizontal axis perpendicular to the above-mentioned vertical plane counted from $0°$ to $90°$ from the horizontal, which brings the antenna boresight in the direction of the satellite. The elevation angle is obtained from the corresponding family of curves of Figure 2.24 which correspond to the following expression:

$$E = \arctan\left[\frac{\cos\Phi - \dfrac{R_e}{R_e + R_o}}{\sqrt{1 - \cos^2\Phi}}\right] \quad \text{(degrees)} \tag{2.12}$$

where

$\cos\Phi = \cos l \cos L$
R_e = radius of the Earth = 6378 km
R_o = altitude of the satellite = 35 786 km

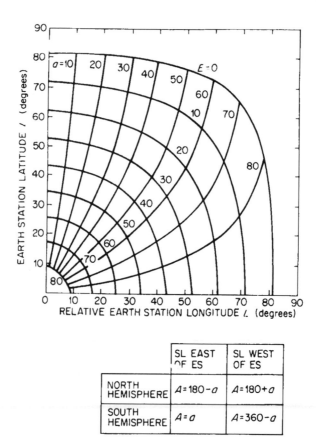

	SL EAST OF ES	SL WEST OF ES
NORTH HEMISPHERE	$A = 180 - a$	$A = 180 + a$
SOUTH HEMISPHERE	$A = a$	$A = 360 - a$

Figure 2.24 Azimuth and elevation angles as a function of the earth station latitude *l* and satellite relative longitude *L*. (Reproduced from [MAR93] by permission of John Wiley & Sons Ltd)

2.3.6   Conjunction of the sun and the satellite

Conjunction of the satellite and the sun at the site of the earth station means that the sun is viewed from the earth station in the same direction as the satellite. As the earth station antenna is pointed towards the satellite, it now becomes also pointed towards the sun. The antenna captures the radio frequency power radiated by the sun and this increases the noise at the antenna noise. The antenna noise increase is discussed in section 3.3.10.

As the satellite rotates along with the Earth, conjunction of the satellite and the sun is a momentary event. It is predictable and actually happens twice per year for several minutes over a period of 5 or 6 days [MAR93, Chapter 7]:

—before the spring equinox and after the autumn equinox for a station in the northern hemisphere:

—after the spring equinox and before the autumn equinox for a station in the southern hemisphere.

2.3.7   Orbit perturbations

Actually, a geostationary satellite does not exist: indeed, Newton's law considers an attracting force exerted on the satellite by a point mass, and oriented towards that point mass. Actually, the Earth is not a point mass, there are other attracting bodies apart from the Earth, and other forces than attraction forces are exerted on the satellite. These effects result in orbit perturbations.

For a geostationary satellite, the major perturbations originate in:

—the Earth neither being a point mass nor being rotationally symmetric: this produces an asymmetry of the gravitational potential;

—the presence of the sun and the moon as other attracting bodies;

—the radiation pressure from the sun, which produces forces on the surfaces of the satellite body facing the sun.

These effects are discussed in detail in [MAR93, Chapter 7]. The practical consequences are summarised below:

—the asymmetry of the gravitational potential generates a *longitudinal drift* of the satellite depending on its station longitude. Actually, there are four equilibrium positions around the Earth where this drift is zero, two of which are stable (at 102° longitude west and 76° longitude east) and two unstable (at 11° longitude west and 164° longitude east). Left to itself, a geostationary satellite would undergo an oscillatory longitudinal drift about the stable positions with a period depending on its initial longitude relative to the nearest point of stable equilibrium. The evolution of the longitude drift with respect to a point of stable equilibrium is shown in Figure 2.25;

—the attraction of the moon and the sun modifies the *inclination of the orbit* at a rate of about 0.8° per year;

—the radiation pressure from the sun creates a force which acts in the direction of the velocity of the satellite on one half of the orbit and in the opposite direction on the other half. In this way the circular orbit of a geostationary satellite tends to become *elliptical*, as illustrated in Figure 2.26.

The ellipticity of the orbit does not increase constantly: with the movement of the Earth about the sun, since the apsidal line of the satellite orbit remains perpendicular to the direction of the sun, the ellipse deforms continuously and the eccentricity remains within limits.

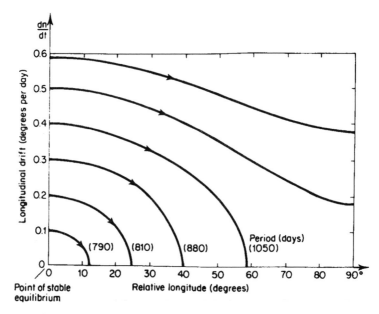

Figure 2.25 Evolution of the longitude drift of a geostationary satellite as a function of the longitude with respect to a point of stable equilibrium. (Reproduced from [MAR93] by permission of John Wiley & Sons Ltd)

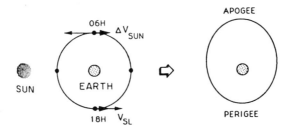

Figure 2.26 Effect of sun radiation pressure on the eccentricity of the orbit. (Reproduced from [MAR93] by permission of John Wiley & Sons Ltd)

2.3.8 Apparent satellite movement

2.3.8.1 *Effect of non-zero inclination*

The track on the Earth of a geostationary satellite with non-zero inclination displays a 'figure of eight' with a 24 hour period, as indicated in Figure 2.27.

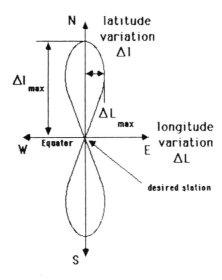

Figure 2.27 Figure of eight as a result of non-zero inclination (24 hour period). (Reproduced from [MAR93] by permission of John Wiley & Sons Ltd)

This can be understood by considering that the satellite is at its nominal position in the equatorial plane when it passes through the nodes of the orbit, then proceeds on a trajectory that is above the equatorial plane from the ascending node to the descending node and below the equatorial plane from the descending node to the ascending node. This explains the north–south component of the 'figure of eight'. The longitudinal component can be understood by observing that the projection B of the satellite on the equatorial plane, as illustrated in Figure 2.28 by the dotted curve, does not have the constant angular velocity of the point

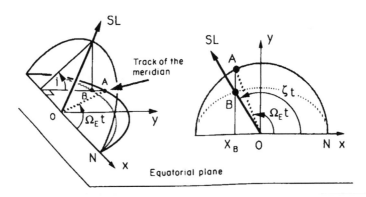

Figure 2.28 Projection of the movement of an earth synchronous satellite in the equatorial plane. (Reproduced from [MAR93] by permission of John Wiley & Sons Ltd)

A which is subjected to the constant angular velocity of the satellite in its orbit. There is consequently an apparent east–west movement of the satellite with respect to the reference meridian on the surface of the Earth (that of the satellite on passing through the nodes).

The latitudinal variation corresponding to the amplitude of the figure of eight above or below the equator is equal to the inclination angle i, and the maximum longitudinal variation with respect to the reference meridian is equal to $4.36 \times 10^{-3} i^2$ (all values in degrees). Therefore for small inclination values, say $0.1°$, the figure of eight can be considered to reduce to a north–south oriented segment, with latitudinal extent from $+i$ degrees to $-i$ degrees.

The transformation of the track of the satellite from a dot on the equator for a perfect geostationary satellite to a north–south oriented segment about this dot for a slightly inclined orbit translates into an apparent movement of the satellite in the sky as viewed from the earth station. This apparent movement leads to a 24 hour period variation of the elevation angle for a station located in the meridian of the nominal satellite meridian, and of the azimuth angle for a station located on the equator east or west of the satellite. For any other station the apparent motion leads to a combined variation of both elevation and azimuth angles.

2.3.8.2 *Effect of non-zero eccentricity*

Figure 2.29 illustrates the effect of non-zero eccentricity: the successive positions of two satellites are represented. One is on a circular orbit, and the other is on an elliptical orbit of the same period.

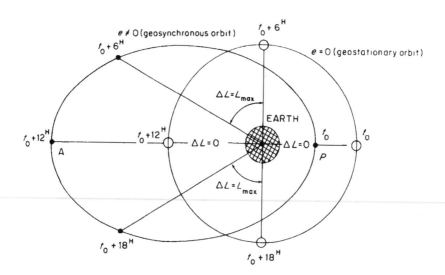

Figure 2.29 Effect of non-zero eccentricity. (Reproduced from [MAR93] by permission of John Wiley & Sons Ltd)

The subsatellite point remains in the equatorial plane and displays a 24 hour period oscillation about its nominal position when the satellite is at perigee or apogee. It can be shown that for small eccentricity values, as considered here, typically less than 0.001, the maximum longitudinal amplitude of the oscillation is 114*e* degrees, where *e* is the value of the eccentricity [MAR93, Chapter 7].

2.3.8.3 *Effect of east–west drift*

The east–west drift of the satellite induces a similar drift of the subsatellite point. This drift is oscillatory and periodic about the nearest point of stable equilibrium. The period and the amplitude of the longitudinal oscillation can be fairly large as indicated by Figure 2.25. Therefore, the apparent movement of the satellite is to be considered as a long term movement compared to the previously discussed movements.

2.3.8.4 *Combined effects*

As a consequence of perturbations, the satellite displays an apparent movement in the sky relative to its nominal position. This apparent movement is the resultant of the combined effects of oscillations of period 24 h due to the non-zero inclination (north–south oriented oscillations) and non-zero eccentricity (east–west oriented oscillations) and the long term drift of the mean longitude (east–west oriented drift). For an earth station, the apparent movement translates into variations of the elevation and azimuth angles with a 24 hour periodic component superimposed on a long term drift. Figure 2.30 gives an example of such variations.

2.3.9 Orbit corrections

2.3.9.1 *Station keeping window*

It is mandatory to take corrective actions to prevent the satellite from straying away from its nominal position in the equator in a given earth meridian. These corrective actions are part of the so-called 'station keeping' procedures. The objective is to maintain the satellite within a station keeping 'window', as represented in Figure 2.31.

The station keeping window is defined by two angles at the earth centre which limit the maximum excursions of the satellite in longitude and latitude: one within the equatorial plane, the other in the satellite meridian. The maximum value of the residual eccentricity determines the overshoot of the radial distance. Therefore the satellite is maintained within the volume represented in Figure 2.31 by the box. The indicated dimensions of the box correspond to a typical window specification of $\pm 0.05°$ in longitude and latitude, and a residual eccentricity of 0.001.

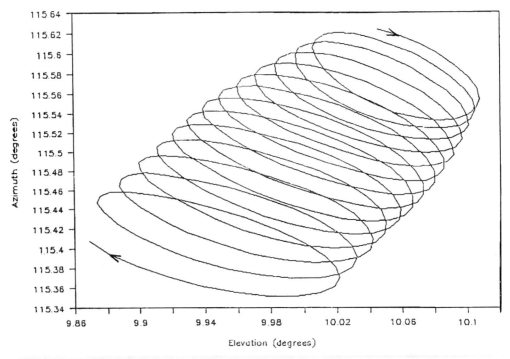

Figure 2.30 Variations of azimuth and elevation angles with time as a result of orbit perturbations. (Reproduced from [MAR93] by permission of John Wiley & Sons Ltd)

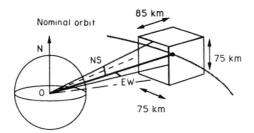

Figure 2.31 Station keeping 'window'. (Reproduced from [MAR93] permission of John Wiley & Sons Ltd.)

2.3.9.2   *Correction procedures*

To maintain the satellite within the box, orbit corrections are achieved by applying velocity impulses to the satellite at a point in the orbit. These impulses are generated by activating the thrusters that are mounted on the satellite as part of the propulsion subsystem. The satellite can be kept at station as long as there is enough propellant left for the thrusts to be produced. When no propellant is left, the satellite drifts in space out of control. This is the end of its operational life. To avoid a possible collision with other geostationary satellites, satellite operators usually keep some amount of propellant for generating a final impulse that positions the satellite into an orbit sufficiently far away from the trajectories followed by its geostationary neighbours.

There is a tendency today, among satellite operators, to relax the north–south specification of the station keeping window. This allows substantial propellant savings, and therefore satellite lifetime extension. However, this results in a higher depointing loss for fixed mounted antennas on the ground (see Chapter 5, Section 5.2).

2.3.10   Doppler effect

The Doppler effect is a change in the receive frequency with respect to the transmit frequency as a result of a non-zero velocity of the transmitter relative to the receiver. If the transmitter transmits at a frequency f, the frequency received by the receiver is $f \pm \Delta f$. The frequency shift Δf is given by:

$$\Delta f = f \frac{V_r}{c} \quad \text{(Hz)} \tag{2.13}$$

where V_r is the absolute value of the relative velocity of the receiver with respect to the transmitter and c is the speed of light ($c = 3 \times 10^8$ m/s).

Communications with geostationary satellites experience a small Doppler effect as a result of the movement of the satellite within its station keeping window. With non-regenerative satellites, the Doppler effect acts twice: once on the uplink, with value Δf_U, and a second time on the downlink with value Δf_D, so the maximum overall frequency shift $\Delta f_{T,max}$ at the receiving earth station is given by:

$$\Delta f_{T,max} = (f_U + f_D) \frac{V_r}{c} \quad \text{(Hz)} \tag{2.14}$$

A typical maximum value for the satellite relative velocity $V_{r,max}$ is 10 km/hour, i.e. 3 m/s, so this generates a maximum frequency shift $\Delta f_{T,max}$ with respect to the transmitted carrier typically equal to 100 Hz at C-band and 260 Hz at Ku-band. This must be accounted for in the design of demodulators, specially at low data rate (a few kb/s), by implementing carrier recovery devices with the ability to track the carrier over the expected frequency span.

2.4   SATELLITES FOR VSAT SERVICES

The selection of satellites for VSAT services entails technical, administrative and commercial aspects.

Table 2.2   Typical values of EIRP and *G/T* for geostationary satellites

Band	Type of converge	EIRP	G/T
C-band	Global beam	24 to 30 dBW	-13 to -8 dBK^{-1}
	Zone beam	30 to 36 dBW	-8 to -3 dBK^{-1}
	Spot beam	36 to 42 dBW	-3 to $+3$ dBK^{-1}
Ku-band	Zone beam	36 to 42 dBW	-7 to -1 dBK^{-1}
	Spot beam	42 to 52 dBW	-1 to $+5$ dBK^{-1}

First the satellite must be positioned at a longitude where it is visible from any earth station in the network. The longitude of existing satellites can be obtained from the ITU, but the documents identify the registered satellites, many of which may not yet be or may never become operational. Apart from the ITU, a number of satellite almanacs are available such as the World Satellite Almanac [LON91] with updates published yearly in the World Satellite Annual. Then the elevation angle should be computed for extreme stations and check be made that it is large enough (10° is a minimum value in open areas).

Second, the satellite coverage should be matched to the network geographical extension. So further investigation should include examination of maps indicating contours at constant EIRP (Effective Isotropic Radiated Power) and *G/T* (receiving figure of merit). As such contours correspond to constant satellite antenna gain, the achieved values depend on the type of coverage: a global coverage offers a worse link performance than zone or spot beam coverage. Candidate satellites should be selected upon the minimum required values of the EIRP and *G/T* which allow the requested link performance.

Third, a check of all station pointing angles (azimuth and elevation) should be performed for all planned sites in such a way that no obstructing obstacles prevent the earth station from accessing the satellite, once installed.

Finally, the selection procedure includes negotiations related to regulatory and financial matters.

Table 2.2 indicates typical values of EIRP and *G/T* for geostationary satellites depending on type of coverage and frequency bands.

3 REGULATORY AND OPERATIONAL ASPECTS

This chapter aims at providing a survey of the main items to be considered when planning to install and operate a VSAT network. The regulatory aspects are considered first. They are important as they must be considered early enough because of the procedural delay. The operational aspects are then discussed, introducing in the first place the user's most obvious concerns, and then the various items that are part of the network life-cycle from installation to testing and carrying out the performance from day to day.

3.1 REGULATORY ASPECTS

Most practical aspects about regulations have been introduced in Chapter 1, section 1.9. Further information is given here for a better perception of the underlying rationale. Regulations apply to different topics:

—licensing of operation

—licensing of equipment

—access to the space segment

—permission for installation

3.1.1 Licensing of operation

The concern reflected here is to ensure compatibility between radio networks by avoiding harmful interference between different systems. By doing so, any licensed operator within a certain frequency band is recognised as not causing unacceptable interference to others, and is protected from interference caused by others.

For that purpose there is a procedure, based mainly on Article 11 and some appendices of the International Telecommunications Union (ITU) regulations:

(a) Application:
   The pertinent earth station data are delivered by the operator, often in forms standardised by the national telecom authority. The telecom authority then registers the earth station data and fills in forms standardised by the ITU. Data are to comply with Appendix 3 of the ITU Radio Regulations. These filled-in forms serve as inputs to the next procedural step, namely coordination.

(b) Coordination:
   One distinguishes between coordination of earth stations against terrestrial services, and coordination of earth stations against other satellite networks. The latter case is rare and will not be penetrated here. In the first case, a protection zone around the earth station, a so-called coordination area, is calculated according to Appendix 28 of the ITU Regulations. A coordination letter, together with the filled-in forms and a plot of the coordination area, is sent to the telecom authorities of the countries affected by that area and to the ITU. Problems arising are to be solved bilaterally. Coordination takes place at a regional level. If the pertinent coordination area does not cover any part of any other country, such a coordination is unnecessary.

(c) Notification:
   When coordination is over, the ITU checks that the procedure has been properly followed. If so, the earth station data are entered into the so-called Master International Frequency Register (MIFR) of the ITU.

3.1.2   Licensing of equipment

The above procedure is workable in a situation where the planning and installation rate of new systems is low. With the increasing demand for installation of small earth stations, the administrative workload has become very burdensome. In some countries, like the USA and Japan, simplified procedures, based on equipment licensing only, have been considered. Equipment standards for VSATs have been elaborated by the national telecommunication authorities or standardisation organisations in those countries. Standards for the USA, Japan and Europe are given in Table 1.11. These standards are largely inspired from the work performed within the ITU by the ITU-R Task Groups.

In the USA, the Federal Communications Committee (FCC), the national telecommunication authority, has issued a policy for a simplified licensing procedure based on type approval of VSAT stations [ITO92] [MIT93]. Therefore, earth stations operating in Ku-band are only required to obtain a licence as a part of a system. It takes about 90 to 120 days to process the licence application. Once the system is licensed other stations with the same parameters in that system can be installed without applying individually for a licence. In addition, if the parameters indicated in Table 1.11 are satisfied, detailed information is not needed for licensing.

In Japan, a similar procedure is used. Once the parameters are checked and certified for an applicant system, a type approval is issued [ITO92]. For VSAT earth stations, licences can be simultaneously granted for multiple stations. A licensed radio operating engineer need not attend each station provided that he can control and monitor the operation of the VSAT station from the hub. The interval of the periodic inspection of the VSAT station by official inspectors is now five years, compared to one year for a conventional earth station [FUJ93].

In Europe, individual countries have different policies. Type approval is not yet effective except for a few countries and specific services. It is believed that the work of ETSI on standardisation will provide useful guidelines for national telecommunication authorities to lead the way towards a pan-European simplified licensing procedure [SAL93]. A recent encouraging fact is the vote by the European Parliament in 1993 of a new regulation allowing a small earth station approved for use in a country belonging to the European Union to be sold everywhere within this Union. This should make it easier and cheaper to install VSAT networks across Europe.

3.1.3   Access to the space segment

Concerning the access to a satellite transponder, the considered national telecommunications authority, if a signatory to an international organisation such as INTELSAT or EUTELSAT, has the responsibility of delivering the space capacity of the satellite owned by these organisations. Should another spacecraft, not belonging to one of these organisations, be considered for use by the applicant VSAT network operator, then the national authority, as a signatory, has to consult the above organisations. In any case, the applicant operator of a VSAT network is compelled to fulfil the requirements imposed by the satellite operator in terms of earth station maximum EIRP, G/T, frequency stability and control of transmission.

3.1.4   Permission for installation

Installation of a VSAT encompasses problems relating to planning or zoning controls, building and personal safety. The VSAT should comply to local regulations dealing with environmental protection. Finally, landlord permission to dig for cable ducts or install roof mounted antennas should be treated as contractual matters between the landlord and the tenant.

3.2   INSTALLATION

3.2.1   Hub

As the hub is relatively large, the installation of it is relatively complex and expensive. Civil works may be necessary. Typically, it takes between one and four

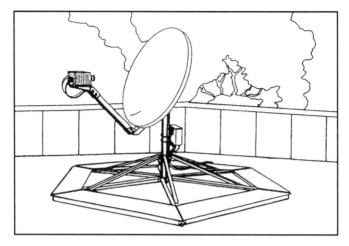

Figure 3.1 Non penetrating roof mount for VSAT [JON88]. Reproduced by permission of Communications News

weeks to install a hub station depending on its size and the selected site. This does not include on-site testing of the equipment.

3.2.2   VSAT

The major problem with VSAT installation is that it involves potentially hundreds of remote VSAT locations with a very wide variety of users, landlords, site conditions, and local zoning requirements.

The VSAT can be roof mounted (Figure 3.1), wall mounted (Figure 3.2) or ground mounted. When ground mounted, the VSAT should be secured with fences (Figure 3.3) to prevent people or animals from getting hurt or damaging the outdoor unit. However, fences are not a strong protection against vandalism.

A typical VSAT installation generally requires three visits to each site [ICO93]: site survey, basic site preparation, equipment installation and test. About 20% of the sites will require an installation revisit.

The following figures may be used for installation planning: the average VSAT cable run is 60 metres; 30% of VSATs are ground mounted, 70% roof mounted. Possibly 15% of roof mounts require special engineering. Fences are recommended for all ground mounts.

Personal computer quality primary power should be available on the site.

3.2.3   Antenna pointing

Once the equipment has been installed, the antenna must be pointed towards the proper satellite. The antenna is oriented according to the azimuth angle Az and elevation angle E, whose expressions are given in Chapter 2, section 2.3.5. These expressions can be used for a coarse orientation of the antenna. The azimuth angle is defined from the geographic north while the magnetic north is given by

Figure 3.2 Wall mount for VSAT [JON88]. Reproduced by permission of Communications News

Figure 3.3 VSAT rounded by fences on the user's premises [SAL 88]

a compass used on the site. The difference is the magnetic declination whose value depends on the site location and the year. The elevation angle must be measured from the horizon, which is defined by the local horizontal plane and easily determined from a spirit level.

Once coarse orientation has been achieved, a more precise orientation is performed according to maximisation of the power received from a satellite beacon or a downlink carrier. For a large hub equipped with a tracking antenna, the tracking equipment can be activated and the orientation of the antenna will remain in the direction of the satellite within the precision of the tracking equipment whatever the subsequent motion of the satellite within its station keeping window. The tracking error is of the order of $0.2\theta_{3dB}$, where θ_{3dB} is the half-power beamwidth of the earth station antenna (see Appendix 4 for definition). Small hub stations and VSATs are not equipped with a tracking antenna and the orientation of the antenna will remain at its initial pointing, assuming that no severe, hazardous force is exerted on the antenna equipment (e.g. from strong wind). Any subsequent motion of the satellite translates into a depointing angle, and the corresponding loss of gain has to be accounted for in the link margin. The maximum gain loss then depends on the initial pointing error and the limits of the satellite motion. This is discussed in more detail in Chapter 5, section 5.2.

3.3   THE CUSTOMER'S CONCERNS

A VSAT network most often replaces an existing leased line data network. The results of a US user survey [JOH 92] indicate that the reasons for using VSAT services are, in order: cost savings (91%), flexibility (84%), reliability (80%), data rates supported (65%), no other services meet needs (41%). This section attempts to list some aspects to be looked at when considering VSAT technology.

3.3.1   Interfaces to end equipment

The indoor unit (IDU) is the part of the network most visible to the user, as it is most often installed in his own office. The IDU is the terminating equipment of the VSAT network, to which the user connects his own terminals. The IDU incorporates a number of input/output ports with specific connectors which must be compatible with that of the user's terminal.

   With data networks, the customer wants to be able to use satellite channels and VSATs in a manner which is transparent to existing and future applications. Often the customer is interested in replacing an existing network but he is usually not willing to replace current equipment such as cluster controllers, front end processors, or other data concentration devices, nor to change the interfaces to that equipment. A customer may be reluctant even to reconfigure the equipment by changing device addresses or the duration of timers [EVE92, pp. 156–157]. Therefore, it is important that all physical interfaces be software defined and downline loadable from the Network Management System (NMS) located at the hub station. Modifications to individual operational interfaces, within a VSAT, should not affect other operational interfaces at the same location.

3.3.2   Independence from vendor

The general functions of a VSAT network, as discussed in Chapter 1, are the same across all vendor products. However, each VSAT has a proprietary design and proprietary protocols. Therefore, VSAT equipment from different vendors cannot share the same satellite channels nor the same network hub equipment in the case of a star network [EVE92, p. 161].

3.3.3   Set-up time

This topic encompasses two aspects:

—the time required to set up the network in a given initial configuration: typically, it takes about 90 days to implement a 100-node network;

—the time to expand the network by addition of new sites: a VSAT can be added within a few days. This compares favourably with several weeks' waiting time for the installation of a terrestrial leased line.

With satellite news gathering (SNG), the VSAT can be installed and in operation typically within 20 minutes.

3.3.4   Access to the service

Many VSAT networks are initially one-way networks used, for instance, for broadcasting of video. Most often, the customer then wishes to upgrade the service into a two-way network for data transmission. Alternatively, broadcast video is a cheap option once the network is installed for data transmission.

It may be worthwhile for the network operator to ask the network provider that installation tests be performed prior to full deployment of the network. This is an opportunity for equipment testing, and also checking that the requested service is actually offered over the subnetwork under test. At this point, the network provider can proceed to traffic measurements and check that the actual traffic is in conformity with design assumptions. Moreover, should the client not be satisfied, the network can be turned down at little expense compared to the cost of turning down the full network.

3.3.5   Flexibility

One of the main advantages of VSAT networks is that network expansion, addition of new terminals and provision of new services can be accommodated without reconfiguring or impacting the operation of the rest of the network.

However, the performance of the network and hence the quality of the service delivered to the user are sensitive to the amount of traffic which increases as more and more terminals of VSATs are added to the network. It is therefore important to allow spare capacity in the space segment and the hub, typically 20% more traffic and 20% more VSATs than initially expected. Growth beyond initial capacity must be orderly and modular. Considering that frequent acquisitions and corporate restructuring are part of today's business world, it is important that the customer not be constrained on its potential growth in telecommunication needs.

3.3.6 Failure and disaster recovery

As telecommunications are a sensitive part of a company's ability to support its business, the customer is concerned with the general feeling that satellite communications are risky by nature, as the network operation relies on a single satellite far away in space without possibility of repair. Most company managers have very little natural confidence in this telecommunications technology which is unknown to them. It is therefore important to establish adequate failure monitoring and diagnostic facilities, restoration procedures, and consistent disaster recovery scenarios. The disaster scenario should be adapted to the customer's particular requirements.

Actions should include the following:

—Hub restoration

—VSAT restoration

—Satellite backup

—Backup terrestrial connections

A hub failure may affect only some of the hub functions and still allow a reduced capability in networking. Should the hub fail, or be destroyed, to a point where the network suffers a complete breakdown, one may consider a properly equipped fixed or transportable earth station to resume immediate operations with no changes to either the satellite or the VSATs. The shared hub option, presented in Chapter 1, section 1.5.5, with its land line connections to the host site, may be more prone to disaster. Further, with many users clamouring for service, priority of restoring could be a problem at the shared hub. The shared hub operator must have a sound, tested plan for such an occurrence.

The network management system (NMS) should perform a centralised failure identification and diagnostic functions at each VSAT. Failure to the card level should be detectable by the NMS. The failure of a VSAT station implies an event which cannot be rectified by commands and the downloading of parameters from the NMS. The successful handling of failures requires accurate and timely detection of the fault. The inclusion of built-in test equipment (BITE) in the VSAT station is essential to support this monitoring facility [EVE92, p. 202].

In case the failure threatens the network integrity, for example if the impaired VSAT transmission would generate interference to other links, there should be immediate termination of transmission from that terminal. A solution is to implement a continuous hub signal which is monitored by the VSAT. The VSAT automatically stops its transmission when not receiving the hub signal.

Satellite failures are rare, but over the typical 15 years of lifetime of a satellite, one must be prepared to face some kind of failure. Satellite depointing is the most probable event and results in a dramatic network breakdown. However, it normally takes no more than a few hours to bring back the satellite to normal operation, and the networking interruption is accounted for in the network availability.

Transponder failure requires shifting the network to another transponder on the same satellite. This possibility is highly dependent on the contractual conditions between the satellite operator and the network operator: the satellite capacity may be leased either as non-preemptible or preemptible. Non-preemptible lease means that the satellite operator warrants the use of the transponder bandwidth and commits himself to do his best to offer the same bandwidth on another transponder in case of failure of the leased one. Preemptible means that the leased capacity cannot be guaranteed over time, and the network operator may be asked to give back the used bandwidth on request from the satellite operator.

Migrating to another transponder on the same satellite implies changing the operating frequencies or polarisation of the entire network. This must be planned in advance so that in case of signal loss for a predetermined time, the VSAT could automatically tune to a new frequency and a new polarisation plane and search for signals from the hub. It should be possible to download the back-up assignment of frequencies and polarisation from the network management system (NMS), to take into account possible updating of the backup transponder scenario.

Finally, there exists a possibility for total satellite failure and subsequent necessity to migrate to another satellite. This implies repointing all remote VSAT antennas, which can be done manually, but takes time, especially with large networks, or automatically, but at a higher cost per VSAT.

In any case, partial or complete breakdown of the network can be avoided if backup terrestrial connections are available. Should a link be disrupted, the traffic on that link can be automatically routed by automatic dial-up modems to a public, either circuit or packet switched, terrestrial network. Figure 3.4 shows a possible implementation for remote-to-host backup interconnections.

The sensing of link failure and automatic recovery via the terrestrial network increase the service availability. Vendors usually offer such features.

3.3.7 Blocking probability

Blocking probability is considered in relation to demand assignment operation, when the total number of VSATs registered in the network possibly generates a traffic demand that exceeds the capacity of the network. When a station needs to establish a connection with another or with the hub, it initiates a request to the

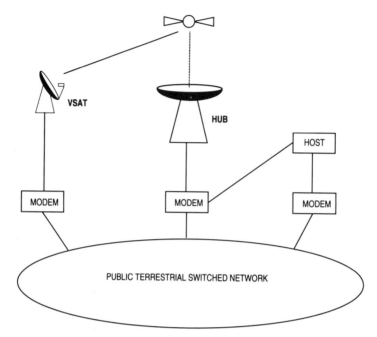

Figure 3.4 Implementation of remote-to-host backup interconnections.

network management system (NMS), and this request is satisfied only if capacity is available. If not, the call is blocked. Chapter 4, section 4.3 gives means for determining the blocking probability. For VSAT networks it typically amounts to 0.1%.

3.3.8 Response time

Response time is defined as the time elapsed between emission of speech and reception of the other talker's response in case of voice telephony communications, or time elapsed between transmission of an enquiry message initiated by pressing the return key of the computer board and the appearance of the first character of the response message on the computer screen.

Response time for data transfer builds up from several components:

—queuing time at the transmitting side as a result of possible delay for capacity reservation before transmission occurs;

—time for transmission of the emitted message which depends on the length of the message and the transmission bit rate;

—propagation time which depends on the network architecture and the number of satellite hops: for a single hop, the propagation time is 0.25 s, and for a double hop 0.5 s. This propagation time occurs on the on-going link from transmitter to receiver and on the return link;

—processing time of the enquiry message at the receiver, and time necessary for generating and transmitting the response;

—protocol induced delay, as a result of error recovery and flow control between the transmitting site and the receiving one.

The VSAT network is only responsible for the routeing delay which includes propagation delay and processing delay as a result of protocol handshake between VSATs and hub front end processor, but excludes the processing delay of the data terminal equipment.

A more detailed analysis of the origin of network delays is given in Chapter 4, section 4.6.

Contrary to a well-established belief, a VSAT network will very likely result in much better response time than the typical private line network. The only physical limitation is the 0.5 second round trip satellite transit delay. The issue then becomes one of cost: how short does the response time really need to be?

3.3.9 Link quality

As mentioned in Chapter 1, section 1.5.1, the user is concerned only by the baseband link quality which is stipulated either in signal power to noise power ratio, S/N, for analogue signals, or in Bit Error Rate (BER) for digital signals. Television can be transmitted in either an analogue or a digital format. Satellite News Gathering tends to use analogue transmission, while business TV is mostly digital. On VSAT networks voice is transmitted in a digital format after being processed in a vocoder. Data, of course, are always digital.

For analogue television, a typical S/N objective is 50 dB. This ensures recovery of a signal with quality suitable for subsequent terrestrial broadcasting or cable distribution. For digital transmission, a typical objective is 10^{-7}. This objective guarantees an acceptable quality for voice or video communications. For data communications, the bit error rate is not a significant parameter, as the transmission can be made error-free thanks to the retransmission protocols that are usually implemented between end-to-end terminals. However, the bit error rate influences the number of required retransmissions, hence influences the delay.

As a result of the symmetry of all links, VSAT networks provide the same service quality to each user. This may not be the case for terrestrial networks.

3.3.10 Availability

In general terms, availability is defined as the ratio of the time a unit is properly functioning to the total time of usage:

$$A(\%) = 100 \; \frac{\text{total usage time-down time}}{\text{total usage time}} \qquad (3.1)$$

Table 3.1 Typical figures for availability

Equipment	Availability (%)
Remote VSAT	99.9
Space segment	99.95
Link	99.9
Hub	99.999
Network	99.7

Network link availability is the percentage of time the service is delivered at a given site with the requested quality (bit error rate less than specified value, for instance one part in 10^7; response time within specified limits, for instance less than 5 seconds). Network availability builds upon equipment reliability, propagation impairments, and sun outage.

More precisely, network availability can be expressed as:

$$A_{net} = A_{Tx} A_{sat} A_{link} A_{Rx} \tag{3.2}$$

where A_{Tx} is the transmitting earth station availability, A_{sat} the space segment availability, A_{link} the link availability, and A_{Rx} the receiving station availability.

Table 3.1 gives some typical figures.

A 99.7% availability corresponds to a cumulated down-time of 26 hours per year. However, it is likely that the user will not accept a service interruption lasting more than typically four hours in a row. Should the service interruption be caused by equipment failure, an appropriate maintenance procedure should be implemented to restore the service within the requested time. Should propagation impairments be responsible for the service interruption, then site diversity can be considered. Finally, backup terrestrial connections may be a means to achieve service continuity.

3.3.10.1   *Earth station availability (transmit or receive)*

There are two aspects to the topic: (a) equipment failure; (b) antenna depointing.

Equipment failure.   A typical mean time between failures (MTBF) for an earth station is 10000 hours (1.15 years). The availability of a remote VSAT station depends on the total repair time. This depends on how easy it is to access the equipment. Spare parts are usually easy to get. Typically, the repair time is from a few hours to a few days. Hence, availability per remote VSAT is typically 99.9% (9 hours/year down time). For a hub station where there is built-in redundancy, the equipment availability is higher, typically 99.999 % (5 minutes per year down time).

Antenna depointing. This may happen as a result of severe mechanical constraints on the antenna reflector, resulting from meteorological events such as strong winds (storms, hurricanes) or heavy snowfall, where wet snow or ice accumulates in the dish. Deicing devices reduce the risk.

3.3.10.2 Space segment availability

There are two aspects to the topic:

—availability of capacity for coping with traffic growth or unexpected demand for a variety of services

—availability of other transponders should a transponder or the entire satellite fail.

Availability of capacity in case of traffic growth. The network operator should be aware of any available satellite capacity in case he needs to expand. It may be a good practice to have spare capacity on the same satellite for occasional video, once a data network is operating.

Availability of capacity depends on the specific region of the world, and varies with time. Truly, adapting the space segment capacity to the demand is a severe challenge for satellite operators, as estimating the demand and scheduling satellite launches to meet the demand entails predictions over a period of 10–20 years, with the uncertainties associated with the spacecraft launching schedule and success. Some regions such as the Pacific region are presently lacking capacity, while there is excess of capacity over North America and Europe. But it has not always been so.

Transponder failure. Should the failed transponder be a non-preemptible one, the network can be transferred within hours to another transponder. The deal is risk versus cost: the network operator may want to lease a non-preemptible transponder with the warranty of being given capacity on another transponder in case of failure. But this costs more than a preemptible transponder with no warranty at all. He then faces the risk of being asked to disembark from his working transponder to leave room for a customer on a failed non-preemptible one.

Satellite failure. Should the entire satellite become useless, then the network must be transferred to another satellite. This implies that one is available in the same band, and with spare capacity. Transferring to this satellite then obliges each site to re-point each antenna. This could take days to weeks, depending on the number of sites. Alternatively, VSATs can be equipped with microprocessor controlled, locally activated automatic repositioning mechanisms. Pointing of the antenna then should be controlled both locally and from the hub station.

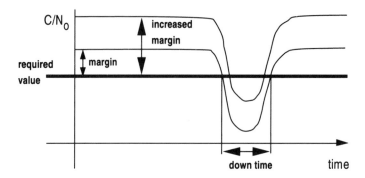

Figure 3.5   Link outage as produced by variation of C/No with rain and sun transit.

3.3.10.3   *Link availability*

Availability requires that the link performance in terms of carrier power to noise power spectral density, C/N_0, be larger than a given value for the considered percentage of time. C/N_0 varies according to propagation effects (mainly rain fade) and sun transit (increase in noise). These effects tend to decrease C/N_0 below its required value and cause link outage as illustrated in Figure 3.5.

Given a specific margin, C/N_0 decreases with rain and sun transit, and may become lower than the required value. C/N_0 resumes its value with margin when rain ceases or sun transit is over. The downtime is smaller as the margin increases; hence, a larger margin value leads to a higher link availability.

Rain fade.   Rain is an issue at Ku-band and even more at Ka-band. It has little effect at C-band. A quantitative discussion of rain effects is presented in Chapter 5.

For voice and video services, reduction of C/N_0 translates into reduction of baseband signal quality (increase in bit error rate, or reduction of baseband signal power to noise power ratio). In data networks, as a result of the end-to-end protocol for error correction, rain fade results in a gradually increasing response time and not a precipitous failure.

An effective way to combat rain attenuation and reduce link outage probability is site diversity [MAR93, Chapter 2]: this means routeing information by either of two stations connected by terrestrial lines, depending on which station is less affected by rain attenuation. The two stations must be sufficiently geographically separated in order not to endure rain simultaneously. As large rain attenuation is caused mostly by storms with limited extension, typically ten kilometres, site diversity is quite feasible with VSAT networks, selecting an appropriate nearby VSAT station as a backup to the failed one. Some systems provide as an option automatic dial-up in case of short term outage for re-routing of the on-going connection to the diverse VSAT, via a terrestrial network which will route data from the failed VSAT to the hub station. Service is automatically restored when the failed VSAT is returned to service.

Table 3.2 Peak antenna noise temperatures at sun transit for various antenna sizes [MOH88]

Antenna diameter (m)	Effective rise in sky noise (K)
1.2	651
1.8	1465
2.4	2605
3.7	5862
9	6012

Sun transit. Sun transit occurs when the conjunction of the satellite and the sun is effective at the site of the earth station. Then the sun passes directly in the line of sight path between the earth station antenna and the satellite, hence the name 'sun transit'.

Conjunction of the satellite and the sun has been introduced in Chapter 2, section 2.3.6. This happens twice per year for several minutes over a period of 5 or 6 days:

—before the spring equinox and after the autumn equinox for a station in the northern hemisphere;

—after the spring equinox and before the autumn equinox for a station in the southern hemisphere.

The sun radiation enters the earth station receiving antenna and increases its noise-temperature. This results in reducing C/N_0. The reduction can be calculated using the equations given in Chapter 5, section 5.3.4.

Table 3.2 gives values of peak antenna noise temperatures at sun transit for various antenna sizes at Ku-band [MOH88]. When no sun transit effect is present the typical antenna noise temperature at Ku-band is in the range 40 to 60 K (see Chapter 5, section 5.3).

The ability of the link to operate through peak sun transit times depends on the built-in link margin. The margin usually implemented for reducing the down time due to rain effects to the typical desired level of link availability of 99.9% is usually large enough to protect the link from sun transit effects. Therefore, the sun transit has little impact on the link availability.

At C-band, the sun transit effect is less pronounced than at Ku-band.

3.3.11 Maintenance

The maintenance concerns the equipment on the ground: the hub station and the VSATs. The customer may wish to be responsible for all of the maintenance, or have the vendor handle some or all of the work.

Maintenance at a shared hub is normally the responsibility of the hub service provider. For a dedicated hub, the network operator may wish to contract out or to perform the maintenance on his own. In terms of maintenance staff, two different categories are required: radio frequency and data communications.

A VSAT station should require as little maintenance as possible as the operational cost of maintenance over a large number of sites scattered over a large service zone would hamper the operational cost of the network. Therefore, it is highly desirable that the maintenance of the VSAT be performed by local people in charge of other duties. For instance, the local technician who maintains the existing PC network on the site can also perform the normal maintenance of the VSAT station. Self diagnostics and box level repair make his task much simpler. In case of extensive trouble shooting, he can call the equipment vendor for *ad hoc* assistance.

The network provider should warrant the availability of the provided hardware and software for a given period, typically two years. He should present a plan as to how all hardware and software will be supported for a minimum of ten years.

3.3.12   Hazards

VSATs are usually located in urban areas or near areas where people and animals may be present. They are usually unattended. Problems are as follows.

3.3.12.1   *Protection of people and animals from radiation*

The electromagnetic radiation should be kept to a harmless low level: typically not more than $10 \, mW/cm^2$ per 6 hours [MOR88]. ETSI's ETS 300 159 specifies that a warning notice should be posted indicating regions where the radiation may exceed $10 \, W/m^2$ ($= 1 \, mW/cm^2$).

3.3.12.2   *Protection of hardware against ill-intentioned people*

Fences are a solution (see Figure 3.3), but it is safer to prevent the outdoor equipment from being not easily accessible, although this renders maintenance more difficult.

3.3.13   Cost

The cost of a VSAT per month per site has been shown to be dependent on the number of VSATs in the network (see Chapter 1, section 1.8), and the cost of the space segment is a sensitive issue. Unfortunately, in most regions of the world the network operator has little freedom of selecting the satellite operator to contract with. The requested availability also has a strong impact on the network cost, therefore the user should not ask for tight specifications unless strictly needed.

One advantage often advocated by VSAT network operators is the control of communications cost. To the initial investment cost is added the maintenance costs; both can be under the control of the network operator.

Therefore, cost containment is a fact. As mentioned above, a company most often turns to VSAT technology in replacement of existing leased lines. As the cost of a link in a VSAT network is not distance sensitive, both immediate and long term cost savings compared to terrestrial alternatives will result if the company encompasses a large number of dispersed sites to be connected.

3.4   VSAT AND HUB EQUIPMENTS

Chapter 1, section 1.6 has presented the architectures of the VSAT station and the hub with their functional split into two parts: the indoor unit (IDU) and the outdoor unit (ODU).

The indoor unit can be considered as the interface to the user terminal, and the outdoor unit as the interface to the space segment.

Tables 3.3 and 3.4 display typical values for the ODU of a hub station and a VSAT.

Table 3.5 indicates the typical features of the IDU of a VSAT or a hub station.

LNA typical noise temperature of today's VSAT receiver is 50 K at C-band and 120 K at Ku-band. Advances in HEMT FET technology now make possible uncooled LNAs having noise temperatures of 35 K at C-band and 80 K at Ku-band [ALB93].

Coherent modulation schemes such as binary phase shift keying (BPSK) or quaternary phase shift keying (QPSK) are used. For acceptable performance, transmission rate should be higher than 2.4 kb/s, otherwise phase noise becomes a problem. For lower data transmission rate values, phase shift keying is avoided and frequency shift keying (FSK) is used instead. For instance, the NORMAC TSAT equipment which operates in the range 300–1200 b/s for SCADA applications utilises a Trellis coded 4FSK modulation scheme [AMU92].

3.5   NETWORK MANAGEMENT SYSTEM (NMS)

Network management covers both operational and administrative functions.

All such functions are performed by the network management system (NMS) usually located at the hub station as indicated in Figure 1.24. It consists of a minicomputer equipped with its dedicated software and displays. This minicomputer is connected to each VSAT in the network by means of permanent virtual circuits. Management messages are constantly exchanged between the NMS and the VSATs and compete with the normal traffic for network resources.

Table 3.3   Typical values for the ODU parts of a hub station

Transmit frequency band	14.0–14.5 GHz (Ku-band)
	5.925–6.425 GHz (C-band)
Receive frequency band	10.7–12.75 GHz (Ku-band)
	3.625–4.2 GHz (C-band)
Antenna	
Type of antenna	Axisymmetric dual reflector (Cassegrain)
Diameter	2 to 5 m (compact hub)
	5 to 8 m (medium hub)
	8 to 10 m (large hub)
TX/RX isolation	30 dB
Voltage Standing Wave Ratio (VSWR)	1.25:1
Polarisation	Linear orthogonal at Ku-band
	Circular orthogonal at C-band
Polarisation adjustment	$\pm 90°$ for linear polarised antenna
Cross polarisation isolation	35 dB on axis
Sidelobe envelope	$29 - 25 \log \theta$
Azimuth travel	120 degrees
Elevation travel	3 to 90 degrees
Positioning	$0.01°/s$
Tracking	Steptrack at Ku-band if antenna larger than 4 m
Wind speed:	
operation	50 to 70 km/h
survival	180 km/h
Deicing	Electric
Power amplifier	
Output power	3–15 W SSPA at Ku-band
	5–20 W SSPA at C-band
	50–100 TWT at Ku-band
	100–200 TWT at C-band
Power setting	0.5 dB steps
Frequency steps	100 kHz to 500 kHz
Low noise receiver	
Noise temperature	80–120 K at Ku-band
	35–55 K at C-band
Operating temperature	$-30°C$ to $+55°C$

3.5.1   Operational functions

Operational functions relate to the network management and provide the capability to dynamically reconfigure the network by adding or deleting VSAT stations, satellite channels and network interfaces. Operational functions also include

Table 3.4 Typical values for the ODU parts of a VSAT station

Transmit frequency band	14.0–14.5 GHz (Ku-band)
	5.925–6.425 GHz (C-band)
Receive frequency band	10.7–12.75 GHz (Ku-band)
	3.625–4.2 GHz (C-band)
Antenna	
Type of antenna	Offset, single reflector, fixed mount
Diameter	1.8–3.5 m at C-band
	1.2–1.8 m at Ku-band
TX/RX isolation	35 dB
Voltage Standing Wave Ratio (VSWR)	1.3:1
Polarisation	Linear orthogonal at Ku-band
	Circular orthogonal at C-band
Polarisation adjustment	$\pm 90°$ for linear polarised antenna
Cross polarisation isolation	30 dB on axis, 22 dB within 1 dB beamwidth
	17 dB from 1° to 10° off axis
Sidelobe envelope	$29–25 \log \theta$
Azimuth adjustment	160 degrees continuous, with fine adjustment
Elevation travel	3 to 90 degrees
Positioning	Automatic positioning optional
Tracking	None
Wind speed:	
operation	75 to 100 km/h
survival	210 km/h
Deicing	Electric (optional) or passive (hydrophobic coating)
Power amplifier	
Output power	0.5 W to 5 W SSPA at Ku-band
	3–30 W SSPA at C-band
Frequency steps	100 kHz
Low noise receiver	
Noise temperature	80–120 K at Ku-band
	35–55 K at C-band
General characteristics	
Effective Isotropic Radiated Power (EIRP)	44 to 55 dBW at C-band
	43 to 53 dBW at Ku-band
Figure of merit *G/T*	13 to 14 dBK^{-1} at C-band
	19 to 23 dBK^{-1} at Ku-band (clear sky)
	14 to 18 dBK^{-1} at Ku-band (99.99% of time)
Operating temperature	$-30\,°C$ to $+55\,°C$

Table 3.5   Typical values for the IDU parts of a VSAT or hub station

Port data rate	1.2 kb/s to 64 kb/s
Interface types	RS232, RS422, RS449 or V35
User interface protocols	SDLC, BSC, X25, Async X3, X28, X29
Operating temperature	0 °C to +50 °C

monitoring and controlling the performance and status of the hub and each VSAT station, and all associated data ports of the network.

This entails operational management tools which provide real-time assignment and connectivity of VSATs, and management and control of new installations and configurations.

The network control software allows automatic dynamic allocation of capacity to VSATs with bursty interactive traffic and to VSATs that will occasionally be used for stream traffic (see Chapter 4, section 4.3). No operator intervention is required to effect this temporary capacity reallocation.

The NMS notifies the operator in case of capacity saturation which prevents more VSAT users from entering the service.

The NMS also handles all aspects related to alarm and failure diagnosis. In particular, in case of any power interruptions at the VSAT stations, the NMS downloads all the relevant software and system parameters for operation restart.

3.5.2   Administrative functions

Administrative functions deal with inventory of equipment, records of network usage, security and billing.

The NMS keeps an account of the VSAT stations installed and operated, the equipment configuration within the hub and each VSAT station, and the port configuration of each network interface.

This information is available on request by the operator, along with statistical information on traffic, number of failures, average time of data transmission delay, etc. This information can be analysed and printed on a daily, weekly or monthly basis as well as being stored on magnetic tape for future reference. It forms the basis for traffic and trend performance analysis, cost distribution based upon usage, etc.

4 NETWORKING ASPECTS

4.1 NETWORK FUNCTIONS

As mentioned in Chapter 1, VSAT networks usually offer communications service between user terminals. These terminals generate baseband signals that are analogue or digital, predominantly digital.

For signals generated by a source terminal and to be delivered to a destination terminal, the VSAT network must provide the following functions:

—establish a connection between the calling terminal and the called one;

—route the signals from the calling terminal to the called one, although the physical resource offered for the considered connection may be shared by other signals on other connections;

—deliver the information in a reliable manner.

Reliable delivery of *analogue* signals implies delivering the signals within acceptable distortion limits and with a sufficiently high signal to noise power ratio S/N.

Reliable delivery of *data* means that data is accepted at one end of a connection in the same order as it was transmitted at the other end, without loss and without duplicates. This implies four constraints [FAI93]:

—no loss (at least one copy of each part of the information content is delivered);

—no duplication (no more than one copy is delivered);

—first in first out (FIFO) delivery (the different parts of the information content are delivered in the original order);

—the information content is delivered within a reasonable time delay.

It has been indicated in Chapter 1, section 1.3 that VSAT networks could be envisaged to support many different types of traffic. However, the network cannot convey all such different types of traffic in a cost effective way.

Therefore, VSAT networks are optimised for a given set of traffic types, which reflect the dominant service demand from the user, and may offer as an option other types of services, but not as efficiently. Most VSAT networks are optimised for *interactive* exchange of data.

This chapter aims at presenting the characteristics of traffic the network may have to convey for interactive data services, and the relevant techniques used for conveying such traffic.

4.2   SOME DEFINITIONS

4.2.1   Link and connection

A link serves as a physical support in a network for a connection between a sending terminal and a receiving one. The network consists of several links and nodes. Every link has two end nodes: a sending one and a receiving one.

In a VSAT network, one finds:

—radio frequency links (uplinks and downlinks);

—cable links between the outdoor and the indoor units, or between the indoor unit and the user terminal;

—possibly terrestrial lines (microwaves or leased terrestrial lines, or lines as part of a public switching network) between the hub and the customer's central facility.

Some connections are one-way, thus requiring that information only travels in one direction: for such connections *simplex* links can be used. An example of a simplex link is a radio frequency wave.

Other connections require interactivity, and hence two-way flow of information. It may be that the information flow is not simultaneous on both ways, but alternate. The supporting links for such connections are named *half duplex* links. An example is when a given radio frequency bandwidth is used alternately by two receiving and transmitting units on a 'push-to-talk' mode: one unit transmits on the bandwidth for some time, while the other unit operates in the receive mode. Once this is done, the transmitter turns to the receive mode, and the receiver to the transmit mode, and information flows the other way round.

Alternatively, the information must travel both ways simultaneously: the supporting links are then called *full duplex* links. An example is the line from a telephone handset to the Indoor Unit (IDU).

Radio frequency links of VSAT networks are inherently simplex links, but a connection requiring full duplex links can be implemented using two radio frequency links: one for each direction of information flow. In a star shaped network, a duplex link between a given VSAT and the hub is constituted for one part by the inbound link and for the other by the outbound link.

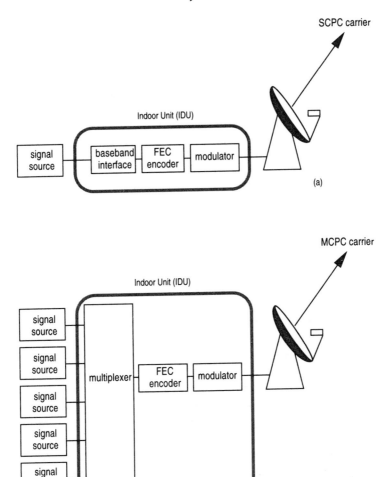

Figure 4.1 (a) Single channel per carrier (SCPC); (b) multiple channels per carrier (MCPC)

A link can support one connection at a time, in a so-called Single Channel Per Carrier (SCPC) mode, or be shared by several connections, in a so-called Multiple Channels Per Carrier (MCPC) mode.

Figure 4.1 illustrates these concepts.

4.2.2 Bit rate

Basically, the bit rate is the number of bits transferred per time unit (second) on a given link. A distinction should be made between:

—the information bit rate R_b, which is the rate at which information bits conveying data messages of interest to the end users are delivered on the link by the data source;

—the channel bit rate R_c, which corresponds to the actual bit rate on a given link while the connection is active. Along with information bits, other bits for error correction and signalling purposes may also be transmitted, so that the channel bit rate on the link is higher than the information bit rate. The channel bit rate imposes bandwidth requirements to the physical support depending on the format used at baseband to represent a bit or a group of bits, also called symbol, and, at radio frequency, on the type of coded modulation used;

—the average bit rate $\langle R \rangle$: links may not be active at all times as connections may be used intermittently, and actually are frequently inactive in case of bursty traffic, made of short data bursts at random intervals. Therefore the average transmitted bit rate is lower than the observed bit rate at times when the link is active. Averaging may apply to either the information bit rate or the channel bit rate.

Consider, for instance, a user terminal acting as a signal source and delivering messages at an average rate of one message per second to a VSAT for transfer to the hub station over a satellite link (Figure 4.1a). Every message contains 1000 information bits. The baseband interface of the indoor unit (IDU) of the VSAT adds some overhead $H = 48$ bits to the message and sends a data unit consisting of the data field $D = 1000$ bits preceded by the overhead $H = 48$ bits to the FEC encoder at a rate $R_b = 64$ kb/s. Therefore the data unit has a duration of 1048/ 64 000 seconds, which is equal to 16.375 ms. The FEC encoder adds one redundant bit to every received bit which means a code rate $\rho = 1/2$. The data unit now modulating the carrier consists of $(D + H)/\rho = 2 \times 1048 = 2096$ bits, and those bits are still occupying a time interval of 16.375 ms corresponding to the duration of the data unit. Thus, the channel bit rate is $R_c = ((D + H)/\rho) \times (1/ 16.375 \text{ ms}) = 128$ kb/s. The average time interval between two messages being 1 second, the average information bit rate $\langle R_b \rangle$ is:

$$\langle R_b \rangle = 1000 \text{ bits}/1 \text{ s} = 1 \text{ kb/s}$$

The link being active at rate $R_c = 128$ kb/s only 16.375 ms out of every second, the average channel bit rate $\langle R_c \rangle$ is:

$$\langle R_c \rangle = R_c \times (16.375 \text{ ms}/1 \text{ s}) = 2.096 \text{ kb/s}$$

4.2.3   Protocol

A protocol is a procedure for establishing and controlling the interchange of information over a network.

For non-data type traffic, the protocols are usually simple and reduce to connection set-up between two end points of a link (TV, voice).

Data communications between the different parts of a network, or between different networks, entail a layered functional architecture which describes how data communications processes are handled. A data protocol is a set of rules for establishing and controlling the exchange of information between peer layers of the network functional architecture.

An example of such a layered architecture is that of the Open Systems Interconnection (OSI) developed by the International Standards Organisation (ISO). This reference model is illustrated in Figure 4.3 and will be discussed in more detail in section 4.4.

4.2.4 Delay

Transfer of information from one user connected to a network to another entails some delay. As mentioned in Chapter 3, section 3.3.8, delay originates from queing time, transmission time, propagation time, processing time, and protocol induced delay.

Delay conditions the network *response time* perceived by the user from the instant he requests a service to the instant the service is performed.

The network response time is highly depending on the type of service considered. For instance:

—for a data transfer service, the response time would be measured as the time elapsed from the instant the first bit of the transmitted data message leaves the sender terminal to the instant the last bit of the message is received at the destination terminal;

—for an interactive data or an enquiry/response service, the response time would be measured as the time elapsed between when the 'enter' key is pressed at the remote terminal and the first character of the response appears on the screen.

Delay is one aspect but delay jitter is also of importance for some applications, such as voice or video transmission. Delay jitter represents the amplitude variation of delay value about its average value, and can be characterised for instance by the value of delay standard deviation.

4.2.5 Throughput

The throughput THRU is the average rate of information bits accepted by the receiving terminal:

$$\text{THRU} = \langle R_b \rangle \quad (\text{b/s}) \tag{4.1}$$

The throughput cannot exceed the information bit rate R_b on the link. It may even be lower than this rate because of overheads, message loss, or source blocking time due to flow control. It is bounded by the maximum throughput, which is a function of the network load. As the source increases its input rate, the actual throughput will grow up to a limit and then remain constant or even deteriorate [FER90].

4.2.6 Channel efficiency

The channel efficiency measures the efficiency of the data transfer by comparing the throughput to the information bit rate R_b on the link.

$$\eta = \frac{\langle R_b \rangle}{R_b} \tag{4.2}$$

4.2.7 Channel utilisation

The channel utilisation is the ratio of the time the connection is used and the sum of the idle time plus the time the connection is used.

$$\text{Channel utilisation} = \frac{\text{service time}}{\text{service time} + \text{idle time}}$$

It identifies with the channel efficiency, η, when no overhead is added to the message.

4.3 TRAFFIC CHARACTERISATION

Traffic characterisation entails different aspects depending on the involved parties and the considered time in the evolution of a network:

4.3.1 Traffic forecasts

This means estimating the type and volume of traffic at peak hour that will be conveyed by the network. Such forecasts should include: traffic breakdown among the different services, variability of the traffic volume and breakdown from site to site, and degree of asymmetry of bidirectional services. This information represents valuable input to the network provider for his network design, prior to any operation, and for the dimensioning of links and interface equipment. Unfortunately, the user is most often incapable of stating a precise activity plan, so it is difficult to make any accurate traffic forecasts. It is less of a problem if measurements can be done on an existing terrestrial network to be replaced by the VSAT network.

4.3.2 Traffic measurements

Measuring the traffic deals with collecting actual values of the traffic flows, in order to provide representative values of the parameters included in the traffic models. This implies a clear perception of which parameters are to be measured, and when and where they are to be measured. Measurements are available only once the network is operational or, prior to its installation, on the existing network it is supposed to replace. There is some risk in basing the dimensioning of a VSAT network on traffic measurements performed on an existing network to be replaced by the VSAT network, as the client's staff may change working and communicating habits once the VSAT network is in operation. Therefore, as mentioned in Chapter 3, section 3.3.4, it is prudent to proceed with such measurements during the installation tests prior to the full deployment of the VSAT network, and to make provision for spare capacity, in case of a higher traffic demand than anticipated. Experience shows that the statistical information provided by a network management system (NMS), indicating for example the number of calls and the volume of messages sent into the network by the user's terminal, may be adequate for network monitoring and billing procedures but is not accurate enough for a proper dimensioning of the network [RES93, p. 51]. Indeed, it does not take into account the actual volume of messages generated in the network as a result of information transfer according to end-to-end or local protocols. Such protocols are responsible for error recovery and flow control, and influence the actual traffic volume in the network and the network throughput.

4.3.3   Traffic source modelling

This involves developing adequate synthetic inputs to the network designer, sufficiently simple to allow mathematical treatment, or to limit the load of the simulation tool, and still sufficiently complex to represent the traffic generated by a source in a realistic manner. Traffic source models should as much as possible include parameters that can be interpreted physically. Examples of popular traffic source models are given in Appendix 1.

Traffic sources can be characterised statistically at call level and burst level.

A call is the means by which a terminal connected to a VSAT in the network indicates its intention to send messages to some other terminal. Some networks offer permanent connections between terminals in the form of leased terrestrial lines. In such circumstances, initiating a call is useless, as a physical path is always available along which the sender can send messages to the destination terminal.

VSAT networks may also offer permanent connections between any two terminals: for this, some bandwidth must be reserved for any carrier between the two VSATs to which the terminals are connected (meshed network) or between the two VSATs and the hub (star network). Most often, this solution is not cost effective, and the required bandwidth will be allocated for the time interval when messages are to be exchanged. Thus, demand assignment is a built-in feature of most VSAT networks. Therefore, before sending messages, a terminal must initiate a call which will be processed by the VSAT network management system (NMS).

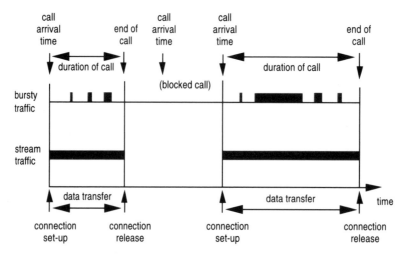

Figure 4.2   Call arrival, connection set-up and data transfer for bursty and stream traffic

Once a connection is established, as a result of call generation and acceptance, the sending terminal is able to transfer messages. Should the message transfer correspond to a continuous flow of data during the call, then the traffic on the connection is of 'stream' type. The characterisation of the traffic during the call (arrival time and duration) has the same parameters as that of the call. Should now the message transfer occur by sequences of small packets, also called bursts, then the traffic is said to be 'bursty', with characteristics of its own.

Figure 4.2 illustrates the above two situations.

4.3.3.1   Call characterisation

Parameters are:

—call generation rate, λ_c (s^{-1})

—mean duration of call, T (s)

When a call is generated a network resource has to be allocated by the network management system (NMS), in the form of a connection over links with the required capacity. The probability of calls being blocked as a result of lack of network capacity can be estimated from the Erlang formula, which assumes that blocked calls are cleared (the NMS does not keep memory of blocked calls). The formula gives the probability that n connections out of C are occupied:

$$E_n(A) = \frac{A^n/n!}{\sum\limits_{k=0}^{k=c}(A^k/k!)} \tag{4.3}$$

where A is the traffic intensity, defined as:

$$A = \lambda_c T \quad \text{(Erlang)} \tag{4.4}$$

and C is the network capacity.

Blocking occurs when $n = C$, therefore the blocking probability is:

$$E_C(A) = \frac{A^C/C!}{\sum\limits_{k=0}^{k=C} (A^k/k!)} \tag{4.5}$$

Formula (4.5) can easily be implemented on a calculator, by using the following iteration:

$$E_n(A) = \frac{AE_{n-1}(A)}{n + AE_{n-1}(A)} \tag{4.6}$$

where $E_o(A) = 1$.

4.3.3.2 Stream traffic

Stream traffic refers to the situation where a continuous transfer of information occurs once the connection between two terminals has been set up for the purpose of that transfer. Therefore, stream traffic can be characterised by the call connection set-up rate λ_c, as this parameter indicates how frequently the traffic is generated by the transmitting terminal. Once the connection is set up, the information transfer is constant and performed at peak bit rate.

An example of such traffic is transfer of video or audio signals. Telephony signals can be considered as stream traffic, although the interactivity between users implies a connection with duplex links, and transfer of information usually is not continuous on each of the two links, as normally one end user would remain silent while the other talks. Therefore, telephony signals, although classified in the stream traffic category, entail some of the characteristics of bursty traffic.

4.3.3.3 Bursty traffic

Bursty traffic refers to intermittent transfer of information during a connection, in the form of individual messages. Messages are short data bursts at random intervals. Typically, this situation arises when a human operated PC is activated by its operator after some thinking time (activation being performed, for instance, by pressing the 'enter' key on the key pad), thus generating the transfer of some text to another terminal. It also results from the specific protocols that are used for data transfer, with information being segmented by the transmitting terminal and segments being acknowledged in the form of short messages by the receiving terminal prior to further transmission by the transmitting terminal.

Bursts introduce new temporal features, characterised as follows:

—the message generation rate, $\lambda(s^{-1})$

—the average length of a message, L(bits)

The interarrival time (IAT) is the time between two successive generations of burst (see Appendix 1). The average interarrival time $\langle IAT \rangle$ is equal to:

$$\langle IAT \rangle = \frac{1}{\lambda} \text{ (s)} \tag{4.7}$$

It is convenient to introduce a measure of how bursty the traffic is. A practical definition for *burstiness*, BU, is the ratio of the peak bit rate, i.e. the rate at which bits are transmitted in burst, to the average bit rate:

$$BU = \frac{R}{\langle R \rangle} = \frac{R}{\lambda L} \tag{4.8}$$

Table 4.1 indicates typical values of the above parameters for different types of services.

Table 4.1   Typical parameter values for examples of stream and bursty traffic

(a) *Stream traffic*

Service	Call generation rate	Average length of message/duration at 64 kb/s	Traffic intensity (Erlang)
Telephony	1 per hour	3 minutes	0.05
Television	1 per day	1 hour	0.042
File transfer (electronic	1 per minute	10^4 bits/0.16 s	0.0026
mail, batch)	1 per day	10^8 bits/1560 s	0.018

(b) *Bursty traffic*

Service	Message generation rate	Average length of message	Burstiness (at 64 kb/s)
Packetised voice	$1\,s^{-1}$	2800 bytes (22 400 bits)	3
Interactive transactions	$0.02–0.2\,s^{-1}$	50–250 bytes (400–2000 bits)	160–8000
Enquiry/response	$0.02–0.2\,s^{-1}$	30–100 bytes (240–800 bits)	400–13 300
Supervisory control and data acquisition (SCADA)	$1\,s^{-1}$	100 bytes (800 bit)	80

4.4 THE OSI REFERENCE MODEL FOR DATA COMMUNICATIONS

This OSI reference model was originally formulated to provide a basis for defining standards for the interconnection of computer systems [TAN89]. Such standards became a necessity when it was experienced that different hardware and software installed in different branches of the same organisation were incapable of exchanging information as a result of incompatibilities. In an attempt to overcome these incompatibilities and create a basis for vendor-independent capabilities of information systems, the ISO has created a model which defines seven functional layers for protocols, as indicated in Figure 4.3.

The figure displays two stacks of layers, one for each of the two interconnected systems. The system on the left is the source machine, generating data to be transmitted in a reliable manner to the system on the right, which is the destination machine. Within one machine, a layer presents an interface consisting of one or more service access points and provides services to the next higher layer while utilising the services provided by the next lower layer. Layers in different stacks at the same level are called 'peer' layers. At every layer, there is a pair of cooperating processes, one on each machine, which exchange messages according to the corresponding layer protocol. The message generated at a given layer is actually passed down to the next lower layer, which is physically implemented by

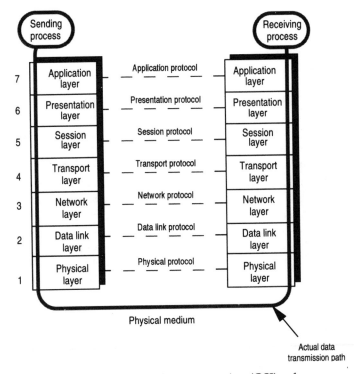

Figure 4.3 The Open Systems Interconnection (OSI) reference model

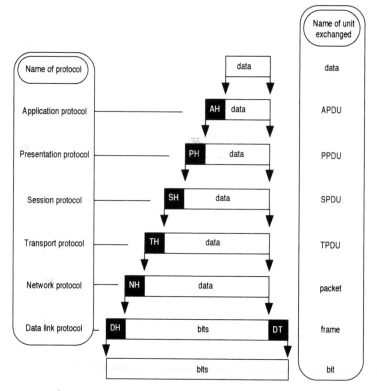

Figure 4.4 Encapsulation from layer to layer in the OSI reference model

hardware and software on the same machine. In this way, the actual data transmission path is down each stack, along the physical medium below the physical layer which connects the two systems, and up the stacks again.

Messages between layers are called Protocol Data Units (PDU). A PDU consists of data preceded by a header (H) and possibly followed by a trailer (T). When a given layer wants to transmit a PDU to its peer layer on the other system, it passes down that PDU to the next lower layer along with some parameters related to the service being requested. Every lower layer accepts the higher layer's PDU as its data, uses the parameters to determine what should be included in the header and appends its own header, and possibly a trailer, so that its peer layer on the other system will know what to do with the data [EVE92, p. 159]. This procedure is called 'encapsulation', and is illustrated in Figure 4.4.

When a message is received in a machine, it passes through the layers. Every layer deciphers its header to derive information on how to handle the data and then strips the header before passing the data up to the next higher layer.

The lower three layers are responsible for the transmission and communications aspects whereas the upper four layers take care of the end-to-end communication and information exchange. A computer system (hardware and software) which conforms to these rules and standards is termed an 'open system'. These systems can be interconnected into an 'open systems environment' with full interoperability.

4.4.1 The physical layer

The physical layer deals with actual transfer of information on the physical medium which constitutes a link, as described in section 4.2.1. Hence, it is concerned with all aspects of bit transmission: bit format, bit rate, bit error rate, forward error correction (FEC) encoding and decoding, modulation and de-modulation, etc.

4.4.2 The data link layer

The data link layer ensures the reliable delivery of data across the physical link. It sends blocks of data called 'frames' and provides the necessary frame identification, error control, and flow control.

4.4.2.1 *Detection of damaged, lost or duplicated frames and error recovery*

The sender organises data in frames of typically a few hundred bytes and transmits the frames sequentially. Frames are identified by means of special bit patterns at the beginning and the end of every frame. Precautions are taken to avoid these bit patterns occurring in the data field.

Upon reception of frames, the receiver sends acknowledgement frames. However, as noise on the link may introduce bit errors, the receiving device must be able to detect such an occurrence. This is performed thanks to checksum bits in the trailer of the frame.

Should the checksum be incorrect, the receiving device sends no acknowledgement frame to the sending device. Not receiving an acknowledgement frame within a given time limit, the sending device retransmits the frame. Hopefully this frame will be correctly received. Otherwise, no acknowledgement is delivered and retransmissions will occur until completion of error free reception.

Multiple transmissions of a frame introduce the possibility of duplicate frames: this would happen, for instance, if the routeing delay exceeds the time limit for retransmission. One or several duplicate frames may be generated before the receiving device has had a chance to transmit its acknowledgement. To obviate this problem, a sequence number in the frame header indicates to the receiver if the received frame is a new frame or a duplicate. Duplicated frames can therefore be discarded.

4.4.2.2   Flow control

A fast sender must be kept from saturating a slow receiver in data. Some traffic regulation must be employed to inform the sender at any instant how much buffer space the receiver has available. This is done by means of sliding window techniques [TAN89, p. 224]: at any instant, the sender maintains a list of consecutive sequence numbers corresponding to frames it is permitted to send. These frames are said to fall within the sending window. Similarly, the receiver also maintains a receiving window corresponding to frames it is permitted to accept.

4.4.3   The network layer

The network layer is responsible for routeing packets from the source to the destination. Therefore, it is concerned with transfer of data over multiple links in the network. This implies identifying the destination (addressing function), identifying the path (routeing), and making sure that the resource is available (congestion control). It also has to identify the link user for purposes of billing (accounting function).

4.4.3.1   Addressing

The network layer is in charge of identifying the destination of data. The receiving device is known by its address. However, this address may be different from one network to the other. For instance, the end terminal may be part of a local area network (LAN) connected to a VSAT. The address of that specific terminal in the LAN may be different from the address of that same terminal in the VSAT network. Therefore, it is up to the network layer to perform the proper address mapping.

4.4.3.2   Routeing of information

The network layer is in charge of determining which links are used. The utilisation of links can be on a fixed assignment basis or on demand.

4.4.3.3   Congestion control

The network layer is also in charge of determining how links are used. For instance it is up to the network layer to regulate the traffic flow in order to avoid congestion on the link, should the traffic flow exceed the capacity of the link.

4.4.3.4   *Accounting*

The network layer supervises the amount of information delivered at any network input and output so as to produce billing information.

4.4.4   The transport layer

The transport layer is in charge of providing reliable data transport from the source machine to the destination machine. Hence, it is an end-to-end layer: it deals with functionalities required between end terminals, possibly communicating through several different networks.

4.4.4.1   *End-to-end transfer of data*

The transport layer accepts data from the session layer, splits it into smaller units if needed, passes these to the network layer and ensures that all pieces arrive correctly at the other end.

4.4.4.2   *Multiplexing*

The transport layer may organise the routeing of several transport connections onto a unique network link. This is called multiplexing and should be transparent to the session layer.

4.4.4.3   *Flow control*

The transport layer is in charge of controlling the flow of information between the end terminals so that a fast terminal does not saturate a slow one. This flow control is distinct from that of the data link layer, although it can be done by similar means.

4.4.5   The upper layers (5 to 7)

These are the session layer, the presentation layer, and the application layer. All are end-to-end layers. These layers are of no concern to VSAT networks. Hence they will not be discussed here. For further information the reader may refer to books on computer communications.

4.5   APPLICATION TO VSAT NETWORKS

4.5.1   Physical and protocol configurations of a VSAT network

A VSAT network essentially provides a connection between any remote user terminal and the host computer. Figure 4.5 illustrates two representations of the end terminals (host computer and user terminals) and the VSAT network in-between. One is a physical configuration which indicates the kind of equipments that support the connection, the other is the protocol configuration which displays

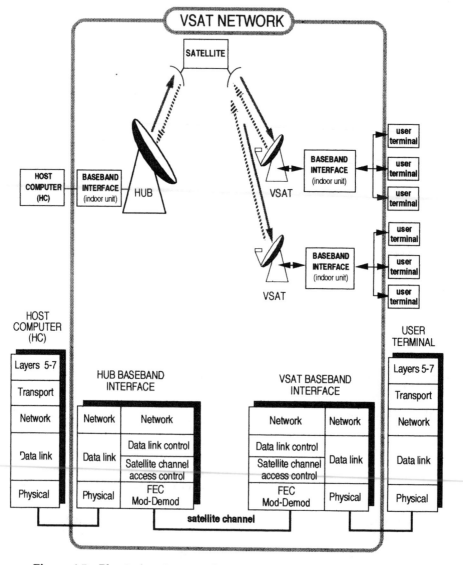

Figure 4.5   Physical and protocol configurations of a VSAT network

the peer layers between the above equipments. The physical configuration shown here displays the hub baseband interface which is part of the indoor unit of the hub, as shown in Figure 1.24, to which the host computer is connected, and the VSAT baseband interface which is part of the VSAT indoor unit, as shown in Figure 1.18, to which the user terminals are connected. The protocol configuration displays the respective stacks of layers from one to seven within the host computer and the user terminal and reduced stacks for the front end processor and the baseband interface of the VSAT indoor unit.

Such a configuration allows for protocol conversion, also named emulation, which will now be presented.

4.5.2   Protocol conversion (emulation)

For the protocol configuration of Figure 4.5, it would be convenient to have a similar one as that of Figure 4.3, where peer-to-peer interactions between end terminals are end-to-end. The VSAT network would act as a pure 'cable in the sky' at the physical layer level, and interconnection of the customer's machines (user terminals and host computer) would be most easy to perform. However, this is not feasible as a result of the characteristics of the satellite channel where information is conveyed at radio frequency with propagation delay and bit error rate. These characteristics differ from those of the terrestrial links for which the protocols used on the customer's machines have been designed. Terrestrial links usually display shorter delays and lower bit error rates than those encountered on satellite links. Consequently, terrestrial oriented protocols may become inefficient over satellite links. Therefore, different protocols must be considered for data transfer over satellite links. Such protocols, however, cannot be end-to-end protocols, as this would imply changing the protocols implemented on the customer's machine, which would be unacceptable to the customer. So, finally, the solution is to implement some form of protocol conversion at both the hub baseband interface and the VSAT baseband interface. The conversion of end terminal protocols into satellite link protocols is called *emulation*, or in a more colloquial manner *spoofing*. Indeed, if the conversion is adequate, that is if it ensures end-to-end transparency, the end terminals will have the impression of being directly interconnected, although they are not.

In Figure 4.5, only the three lower layers (network, data and physical) are emulated. This corresponds to a common situation. However, some services might require that emulation be carried up to the transport layer.

The network layer protocol emulation performs address mapping for the customer's machines. This enables the network addresses to be independent of the customer addresses.

The data link layer is split into two sublayers: the sublayer named 'data link control' provides data link control over the satellite links independently from the data link control between the VSAT network interfaces and the customer's machines. The sub-layer called 'satellite channel access control' is responsible for the access to the satellite channel by multiple carriers transmitted by the VSATs or the hub station. An important aspect here, which is specific to VSAT networks, is that the powered bandwidth of the satellite required for the carrier which provides the connection at the physical level, if allocated on a permanent basis, is poorly used in case of infrequent stream traffic or with bursty traffic. It is, therefore, desirable that this satellite resource be allocated to any VSAT earth station on a demand assignment (DA) basis, as presented in Chapter 1, section 1.5.3, according to the traffic demand and characteristics.

Finally, at the physical level, any earth station (hub or VSAT) has to provide a physical interface which actually supports the physical connection. On the customer's side, the physical interface should be compliant with the customer's equipment hardware. On the satellite side, the physical level should provide protection of data against errors by means of forward error correction (FEC) encoding and decoding techniques, and modulate or demodulate carriers conveying the data.

4.5.3 Reasons for protocol conversion

This section aims at discussing in more detail some of the underlying reasons exposed above.

Satellite links differ from terrestrial links in two ways:

—the large propagation delay, about 270 ms, from one site to another over satellite links in comparison to the much smaller delays encountered on terrestrial networks, typically a few milliseconds to tens of milliseconds;

—the bit error rate: satellite links are corrupted by noise which affects the carrier received by the demodulator. The bit error rate can be reduced to levels typically of 10^{-7} thanks to the use of forward error correction (FEC) but this is still higher than the bit error rate level encountered on terrestrial links.

It will now be shown how these characteristics impact on protocols when used over satellite links.

4.5.3.1   Impact on error control

The following example deals with transmission of a data stream over a connection from the host computer to a user terminal, using automatic repeat request (ARQ) protocols for error control. The data link layer gets its protocol data unit (packet) from the network layer and encapsulates it in a frame by adding its data link header and trailer to it (see Figure 4.4). This frame is then transmitted over the

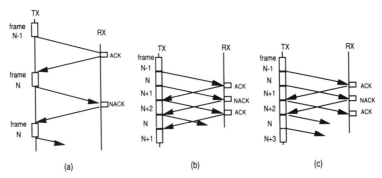

Figure 4.6 (a) Stop-and-wait protocol; (b) go-back-N protocol; (c) selective-repeat protocol

network to the data link layer of the user terminal which verifies the integrity of data by means of the error detection information contained in the trailer of the frame. Should the received data be error free, the user terminal sends a positive acknowledgement (ACK) back to the host computer. If not, it sends a negative acknowledgement (NACK). If the host computer receives no ACK and no NACK after a time-out delay, then it retransmits the frame.

In the following, three ARQ protocols are considered (Figure 4.6):

—a stop-and-wait (SW) protocol: the host computer waits until it receives a positive acknowledgement, ACK, before sending the next frame. If a negative acknowledgement, NACK, is received, the host computer retransmits the same frame (Figure 4.6a);

—a go-back-N protocol (GBN): the host computer transmits frames in sequence as long as it does not receive any negative acknowledgement, NACK. Receiving NACK for frame N, it retransmits frame N and all subsequent frames (Figure 4.6b);

—a selective-repeat (SR) protocol: the host computer transmits frames in sequence as long as it does not receive a negative acknowledgement, NACK. Receiving a NACK for frame N while sending frame $N + n$, it retransmits frame N after frame $N + n$, then continues with frame $N + n + 1$ and the subsequent ones (Figure 4.6c, here $n = 2$).

The three protocols will now be compared on the basis of the channel efficiency. Appendix 2 demonstrates that the channel efficiency η_c for every protocol is equal to:

stop-and-wait: $\eta_{cSW} = D\dfrac{1 - P_f}{R_b T_{RT}}$ (4.9) go-back-N: $\eta_{\text{cGBN}} = D\dfrac{1 - P_f}{L(1 - P_f) + R_b T_{\text{RT}} P_f}$ (4.10)

selective-repeat: $\eta_{\text{cSR}} = D\dfrac{1 - P_f}{L}$ (4.11)

where:

D = number of information bits per frame (bits)
L = length of frame (bits) = $D + H$ (information plus overhead)
P_f = frame error probability = $1 - (1 - \text{BER})^L$, where BER is the bit error rate
R_b = information bit rate over the connection (b/s)
T_{RT} = round trip time (s)

The round trip time (T_{RT}) corresponds to the addition of service times and propagation delays. At time L/R_b, the last bit of the frame has been sent. At time L/R_b plus the propagation delay T_p from the sender to the receiver, the last bit has arrived at the receiver. From a host computer to a user terminal over a *terrestrial link* T_p is about 5 ms. From a VSAT to the hub station over a *satellite link* T_p is about 260 ms (see Figure 2.22). Neglecting the processing time, the receiver is now ready to send the acknowledgement message. Denote by A the acknowledgement frame length and by R_{back} the information bit rate at which the acknowledgement is sent on the return link. At time $L/R_b + T_p + A/R_{\text{back}}$, the last bit of the acknowledgement frame has been sent. At time $L/R_b + T_p + A/R_{\text{back}} + T_p$ the sender has received the acknowledgement. So the round trip time is:

$$T_{\text{RT}} = \frac{L}{R_b} + 2T_p + \frac{A}{R_{\text{back}}} \quad (\text{s})$$ (4.12)

One can neglect A/R_{back} relative to L/R_b as the acknowledgement frame length is much smaller than the frame length L (it often reduces to the header H), and for a VSAT network generally R_{back} is the outbound link bit rate which is usually larger than R_b. Therefore the round trip time can be approximated by:

$$T_{\text{RT}} = \frac{L}{R_b} + 2T_p \quad (\text{s})$$ (4.13)

Figure 4.7 compares channel efficiency η_{cSW} and η_{cGBN} as a function of the round trip time T_{RT} for different values of the bit error rate. The parameter values selected here are:

$D = 1000$ bits
$H = 48$ bits
$L = 1048$ bits
$R_b = 64\,\text{kb/s}$

On a terrestrial link, taking $T_p = 5\,\text{ms}$, T_{RT} would be about 26 ms. On a satellite link, taking $T_p = 260\,\text{ms}$, T_{RT} would be about 536 ms.

With the selective-repeat protocol, the channel efficiency is independent of the round trip time. With the selected parameter values, one obtains the values

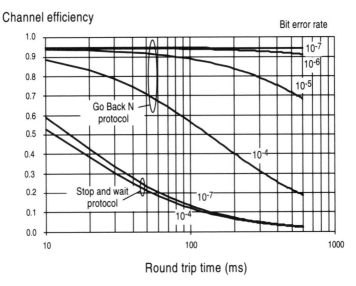

Figure 4.7 Channel efficiency for stop-and-wait, and go-back-N as a function of round trip time T_{RT} and bit error rate (BER)

Table 4.2 Values of channel efficiency η_{cSR} for selective-repeat protocol as a function of bit error rate (BER)

BER	10^{-4}	10^{-5}	10^{-6}	10^{-7}
η_{cSR}	0.86	0.95	0.95	0.95

indicated in Table 4.2. It can be seen that η_{cSR} is always greater than with the two other protocols.

Figure 4.7 indicates that the channel loses much of its efficiency when a stop-and-wait protocol is implemented on a satellite link, as a result of the increased round trip time compared to terrestrial links. The same is true of a go-back-N protocol should the satellite link be of poor quality (bit error rate in the range 10^{-4} to 10^{-5}). If the satellite link has a bit error rate in the order of 10^{-7}, then no degradation is observed. Finally, as seen from Table 4.2, the selective-repeat protocol, which offers a good performance for reasonably low bit error rates, is a good candidate for satellite links as it is not sensitive to a long round trip delay.

4.5.3.2 *Impact on flow control*

Protocols at the data link layer and transport layer level often make use of sliding windows for flow control purposes. In such cases, only positive acknowledgements (ACK) are sent. Not receiving an acknowledgement before a given time-out interval, the sender retransmits the protocol data unit which has not been acknowledged. The sender can only send a limited number of protocol data units following the last acknowledged one. These are said to fall within the sending window. The window slides by one position at every received acknowledgement, therefore initiating the clearance for sending a subsequent protocol data unit. Similarly, the receiver accepts only a limited number of protocol data units before sending a positive acknowledgement. Any incoming protocol data unit that falls outside the window is discarded. The window slides by one position at every emitted acknowledgement, and the receiver can subsequently accept one more protocol data unit.

It can be shown that the channel efficiency for a sliding window protocol with error control based on the selective repeat procedure is [TAN89, p. 243]:

$$\eta_{cSR} = D\frac{(1-P_f)W}{L+R_bT_{RT}} \quad \text{if} \quad W < 1 + \frac{R_bT_{RT}}{L} \tag{4.14}$$

$$\eta_{cSR} \quad = D\frac{1-P_f}{L} \quad \text{if} \quad W \geqslant 1 + \frac{R_bT_{RT}}{L} \tag{4.15}$$

where W is the window size, and the other parameters are as in the previous section.

Figure 4.8 represents the channel efficiency as a function of the round trip time T_{RT} for the example discussed in the previous section. Different window sizes are considered from 1 to 31. The case $W = 1$ corresponds to a stop-and-wait protocol, as presented in the previous section. One can see from the figure and from (4.15)

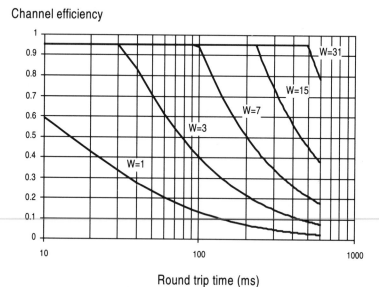

Figure 4.8 Channel efficiency when using a sliding window protocol as a function of round trip time for different window sizes W. The bit error rate on the link is 10^{-7}

that satellite links require large window values to be efficient, while for terrestrial links, small values can be implemented.

This can be explained as follows: the quantity $R_b T_{RT}/2L$ represents the number of protocol data units that the link from the sender to the receiver can hold. The quantity $R_b T_{RT}/L$ conditions the selection of either (4.14) or (4.15) to express the channel efficiency. The horizontal part of the curves of Figure 4.8 are obtained from equation (4.15), while the sharp decreases are expressed by (4.14). The quantity $R_b T_{RT}/L$ represents the total number of protocol data units filling both the direct and return links from sender to receiver. Should the window exceed that quantity, then transmission goes on continuously and the protocol leads to an efficient use of the channel (horizontal part of the curve). If the window is less than that quantity, the sender is blocked when the window is full and has to wait for an acknowledgement to come in before resuming transmission. The channel not being used during this waiting time (the longer the round trip time, the longer the waiting time), the use of the channel is reduced and the channel efficiency decreases.

One can conclude by stating that flow control over satellite links using sliding window protocols is feasible without loss of channel efficiency if the window is large enough.

4.5.3.3 Polling over satellite links

For private networks, IBM SNA protocols or bisynchronous and asynchronous protocols are most often used. Figure 4.9 illustrates a typical IBM SNA synchronous data link control (SDLC) environment, where the host computer communicates with remote cluster control units (CCU) by means of a multidrop line. Data terminals are attached to every CCU by point to point lines. The host

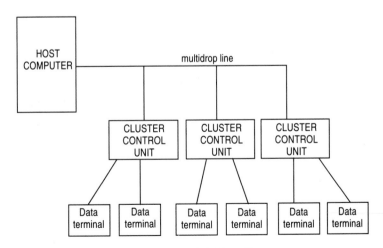

Figure 4.9 Multidrop line

computer manages the transfer of data between itself and the CCUs, on the shared capacity of the multidrop line by means of a technique named polling.

Polling means that the host sends a message to every CCU it controls, enquiring whether or not the CCU has anything to send. Every CCU acknowledges its own poll and sends data if it has data to send. The host then acknowledges reception of data. If the CCU has no data to send, it sends a 'poll reject' message and the host polls the next CCU in a round-robin fashion.

Alternatively, the host when having data to send to a terminal connected to a given CCU sends a select command to the addressed CCU. The CCU sends an acknowledgement to indicate that it is ready to receive the data. The host then

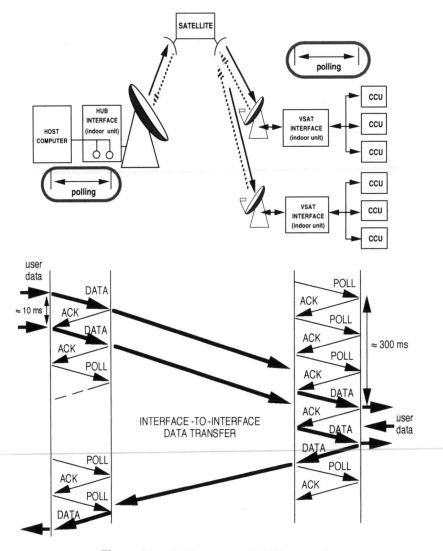

Figure 4.10   Polling over a VSAT network

transmits the data and this is followed by an acknowledgement from the CCU indicating correct reception.

A network would possibly comprise a host and several multidrop lines such as the one illustrated in Figure 4.9, using permanent terrestrial leased lines or temporary connections via the public switched network. It is assumed that the company which utilises the private network wishes to replace part or the whole of it by a VSAT network, as shown in Figure 4.10.

It is undesirable to pass every polling message and its acknowledgement across the satellite. Such a scheme would produce high transmission delays, as every handshake would take 0.5 s, and waste transponder capacity, as every polling is not necessarily followed by transfer of data messages. To avoid these undesirable effects, polling emulation is implemented at the remote sites and at the central facility: at the remote site, the VSAT interface polls the CCUs, thus acting as the host computer in Figure 4.9, while at the central facility, the ports of the hub interface, as many as the remote sites, are polled by the host, acting as 'virtual' CCUs.

Acknowledgements are provided by the hub interface ports as soon as they receive the polling message from the host, and the VSAT interface independently polls every CCU. Data messages are transmitted on the satellite links only if a CCU responds to polling by transmitting data, or if the host selects one of the hub interface ports to transmit data. VSAT and hub interfaces must provide buffering and flow control on the satellite links.

4.5.3.4 *Conclusion*

The above examples show that protocols that perform adequately on terrestrial links may work poorly when used as such on satellite links. Hence, there is a need for protocol tuning or protocol conversion at the interface between end terminals and the VSAT network, or between other networks and the VSAT network.

4.6 MULTIPLE ACCESS

The earth stations of a VSAT network communicate across the satellite by means of modulated carriers. Depending on the network configuration, different types and number of carriers must be routed simultaneously within the same satellite transponder. Figure 4.11 illustrates different possible situations:

—with one-way networks, where the hub broadcasts a time division multiplex of data to many receive-only VSATs, only one carrier is to be relayed by the satellite transponder. Accordingly, there is no other carrier competing for satellite transponder access, and there is no need for any multiple access protocol;

—with two-way star shaped networks, carriers from VSATs and the hub station are competing to access a satellite transponder;

ONE-WAY NETWORKS :

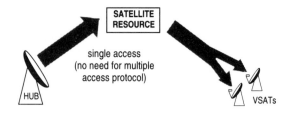

TWO-WAY STAR SHAPED NETWORKS :

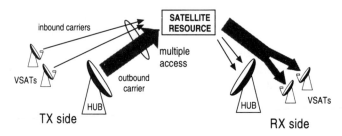

TWO-WAY MESHED NETWORKS :

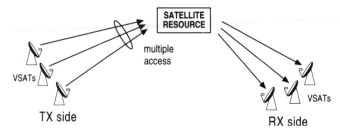

Figure 4.11   Multiple access for different network configurations

—with two-way meshed networks, there is no hub station and the only carriers
competing to access a satellite transponder are those transmitted by the VSAT
stations.

Multiple access is therefore to be considered in the two latter situations only.
Multiple access schemes differ in the way the satellite transponder resource,
which is powered bandwidth during the lifetime of the satellite, is shared among
the contenders.

4.6.1   Basic multiple access protocols

—*Frequency division multiple access (FDMA)* means allocating a given subband of
the transponder to every carrier. The allocated subband must be compatible

with the carrier bandwidth which depends on the bit rate it conveys and the type of modulation and coding (see Chapter 5, section 5.8).

—*Time division multiple access (TDMA)* means allocating the overall bandwidth of the transponder to every carrier in sequence for a limited amount of time, called a time slot.

The sequence may be random, every station transmitting a data packet on a carrier burst with duration equal to a time slot whenever it has data to transmit, without being coordinated with respect to other stations. This is named 'random TDMA' and is best represented by the so-called ALOHA type protocols. As a result of the random nature of transmissions, such multiple access schemes do not protect two or more carrier bursts transmitted by separate stations from possibly colliding within the transponder (that is overlapping in time). The interference which results then prevents the receiving stations from retrieving the data packets from the corrupted bursts. To provide error free transmission, ALOHA protocols make use of ARQ strategies by sending acknowledgements for every packet correctly received: in case of collision, the transmitting stations not receiving any acknowledgement before the end of their time-out interval will retransmit the unacknowledged packet at the end of a random time interval calculated independently at every station, so as to avoid another collision.

Alternatively, the sequence may be synchronised in such a way that bursts occupy assigned non-overlapping time slots. This implies that the time slots be organised within a periodic structure, called a TDMA frame, with as many time slots as active stations (note that the term 'frame', with TDMA, represents a different concept than with computer communications, where a 'frame' is a block of data sent or received by a computer at the data link layer of the OSI reference model of Figure 4.4).

With TDMA, carriers are transmitted in bursts and received in bursts: every burst consists of a header made of two sequences of bits: one for carrier and bit timing acquisition by the receiving VSAT demodulator, another named 'unique word' indicating to the receiver the start of the data field. The header is followed by a data field containing the conveyed data packet. Synchronisation is necessary between earth stations, and the earth station must be equipped with rapid acquisition demodulators in order to limit burst preambles to a minimum.

—*Code division multiple access (CDMA)* is a multiple access technique which does not consider any frequency-time partition: carriers are allowed to be transmitted continuously while occupying the full transponder bandwidth. Therefore interference is inevitable, but is resolved by using spread spectrum transmission techniques based on the generation of high rate chip sequences (or 'code'), one for every transmitted carrier. These sequences should be orthogonal so as to limit interference. Such techniques allow the receiver to reject the received interference and retrieve the wanted message.

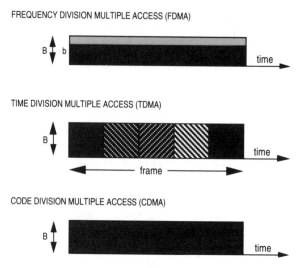

Figure 4.12   Basic multiple access protocols

The selection of a multiple access scheme should take into account the requirement for power and bandwidth not only of the satellite transponder, but also of
the earth stations (VSATs and hub station). Generally speaking, operating a satellite transponder in a multicarrier mode (several carriers sharing the transponder
bandwidth at given time), as with FDMA and CDMA, entails the generation of
intermodulation noise which adds to the thermal noise (see Chapter 5, section
5.4). Carriers conveying a high bit rate are more demanding for bandwidth and
power than smaller carriers. This impacts on the EIRP requirement of the
transmitter: it translates into a higher demand for power from VSAT transmitters
on the inbound links, from the hub station transmitter on the outbound links, and
from the satellite transponder on all links. It also translates into a higher demand
for bandwidth on the satellite transponder.

We will now discuss the practical implementation of these multiple access
schemes in VSAT networks. It will be assumed that a fraction of a satellite
transponder bandwidth is allocated to the VSAT network, hence it may be that the
rest of the transponder is occupied by carriers originating from earth stations
other than those belonging to the considered VSAT network. Indeed, it seldom
happens that the demand for capacity of a VSAT network requires full transponder usage. Therefore, the transponder is actually divided into subbands, every
subband being used by different networks. In a way this represents 'network
FDMA'. This means that the satellite resource available to a given network is only
a fraction of a satellite transponder overall resource, as not only the transponder
bandwidth has to be shared but also the output power. Therefore, a considered
VSAT network benefits neither from the entire transponder effective isotropic
radiated power (EIRP), nor from its full bandwidth.

4.6.2 Meshed networks

The meshed network comprises N VSATs. Every VSAT should be able to establish a link to any other across the satellite.

A first approach is to have every VSAT transmitting as many carriers as there are other VSATs: the information conveyed on every carrier represents the traffic from one to any other VSAT. For permanent full network connectivity, every VSAT should be able to receive at any time all carriers transmitted by the other VSATs in the network.

Figure 4.13 proposes an implementation based on FDMA. Such a configuration requires that every VSAT be equipped with $N-1$ transmitters and $N-1$ receivers. This is costly if N is large, and poses operational difficulties as more transmitters and receivers must be installed at every VSAT each time the network incorporates new VSATs. Moreover, the satellite transponder is occupied by $N(N-1)$ carriers. Such carriers are narrow band ones as they convey low bit rates. This may require frequency stable modulators because guard bands between carriers must be kept to a minimum in order to save satellite bandwidth. As an example, consider a VSAT network with $N=100$ VSATs. The number of transmitters and receivers per VSAT is $N-1=99$. The number of carriers is $N(N-1)=9900$.

A variant to Figure 4.13 is considering the broadcasting capability of the satellite: any carrier uplinked by a VSAT is actually received by all VSATs. Therefore, the overall traffic conveyed by the $N-1$ carriers transmitted by a given VSAT in Figure 4.13 can be multiplexed onto a unique carrier. Receiving that carrier, any VSATs can demodulate it and extract from the baseband multiplex the traffic dedicated to itself. Now, every VSAT still needs $N-1$ receivers, but

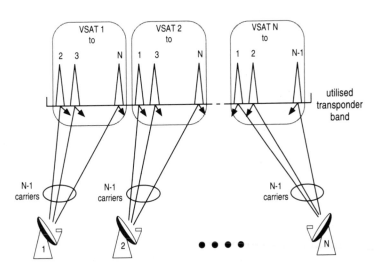

Figure 4.13 Meshed network with N VSATs transmitting as many carriers as there are other VSATs, using Frequency Division Multiple Access (FDMA)

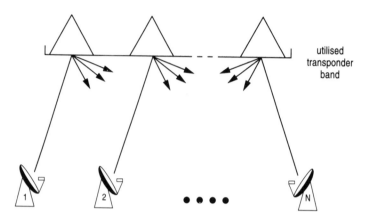

Figure 4.14 Variant of Figure 4.13 where the overall traffic from a VSAT to all other VSATs is multiplexed on a single carrier

only one transmitter. However, the capacity of the transmitted carrier is higher, thus the VSAT transmitter must be more powerful. This scheme is represented in Figure 4.14.

The problem of having many receivers and transmitters comes from the requirement for permanent full connectivity. Actually, there is seldom need for such a requirement: indeed, apart from some broadcasting applications, the customer usually only requests that temporary duplex connections be set up between any two remote terminals attached to two different VSATs in the network. This works out most conveniently through demand assignment (see Chapter 1, section 1.5.3): should a terminal ask for such a connection , then the VSAT it is attached to sends a request on a signalling channel to a traffic control station, which replies by allocating some of the available satellite resource to both the calling and the called VSATs. With FDMA, this resource consists of two subbands on the satellite transponder, one for each carrier transmitted by the two VSATs. So any VSAT needs only to be equipped with one transmitter and one receiver, both tuneable on request to any potential frequency band allocation within the transponder bandwidth.

Should now TDMA be used in Figure 4.14 instead of FDMA, then permanent full connectivity can be achieved with only one carrier being transmitted and received by every VSAT. This looks appealing but one must consider the higher cost of the TDMA equipment, and the fact that permanent full connectivity is not really asked for.

With CDMA, the analysis follows the same lines as with FDMA. With demand assignment, temporary connections are set up by allocating to every transmitting VSAT a specific code. However, there does not seem to be any advantage to using CDMA apart for small VSAT networks operating at C-band. CDMA then offers protection against interference generated by other systems.

Most of today's commercial meshed networks are based on demand assignment FDMA.

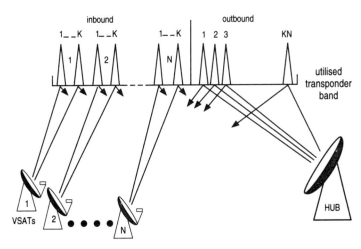

Figure 4.15 Star shaped network with a two-way connection being conveyed on two SCPC carriers: one from the VSAT to the hub station, and one from the hub to the VSAT. Satellite transponder access is FDMA

4.6.3   Star shaped networks

The star shaped network comprises N VSATs and a hub. Every VSAT can transmit up to K carriers, corresponding to connections between terminals attached to the VSAT and the corresponding applications at the host computer connected to the hub station.

4.6.3.1   FDMA–SCPC inbound/FDMA–SCPC outbound

Figure 4.15 illustrates the case where connections between any remote terminal and the corresponding application at the host computer are supported by duplex links, by means of two single channel per carrier (SCPC) carriers: one from the VSAT to the hub station, and one from the hub to the VSAT. Every carrier requires its own modulator and demodulator. Hence, this configuration requires K modulators and demodulators at every VSAT and KN modulators and demodulators at the hub station. This is costly if the number of VSATs is large and K larger than 1. For instance, with $N = 100$ and $K = 3$, three hundred modulators and demodulators are to be installed at the hub.

With demand assignment, frequency agility is required for both transmitting and receiving VSATs.

4.6.3.2   FDMA–SCPC inbound/FDMA–MCPC outbound

Considering that any carrier transmitted by the hub is received by all VSATs, the number of modulators at the hub can be reduced, as indicated in Figure 4.16, by

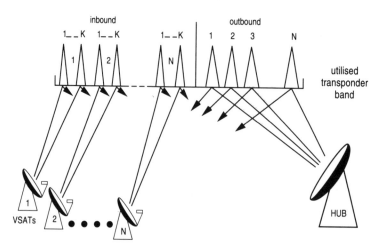

Figure 4.16 Star shaped network with a two-way connection being conveyed on an SCPC carrier from the VSAT to the hub station, and multiplexed with others of the same VSAT on an MCPC carrier from the hub to the VSAT. Satellite transponder access is FDMA

time division multiplexing the traffic from the hub to one VSAT on an outbound MCPC (multiple channels per carrier) carrier. The number of modulators at the hub is now equal to the number N of VSATs. The number of demodulators at every VSAT can be reduced to one.

As the number of multiplexed connections on any outbound carrier may vary with time, the hub modulators and VSAT demodulators must be able to accommodate variable rates. The increased transmitted rate from the hub is higher with MCPC carriers. This translates into a higher demand for power from the hub transmitter.

With demand assignment, frequency agility is required for transmitting VSATs only.

4.6.3.3 FDMA–SCPC inbound/TDM outbound

The number of modulators at the hub and demodulators at the VSATs can even be reduced to one, as illustrated in Figure 4.17, by time division multiplexing all connections from the hub to the VSATs on one MCPC outbound carrier. The hub modulator and the VSAT demodulator can operate at constant bit rate, equal to the maximum capacity of the network. But as a result of the higher bit rate, the demand for power from the hub transmitter is increased. The large imbalance in input power between the low powered inbound carriers and the high powered outbound carrier results in a 'capture effect' at the output of the satellite transponder, when used near saturation: the outbound carrier has a larger share of the output transponder power than its share at the input [MAR93, section 4.4.4, p. 137]. Therefore, less power is available to the inbound carriers.

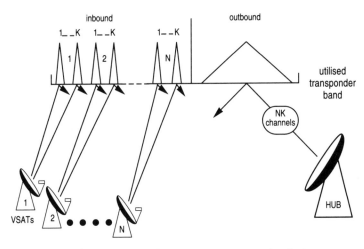

Figure 4.17 Star shaped network with a two-way connection being conveyed on SCPC carriers from the VSAT to the hub station, and multiplexed with all others on the MCPC carrier from the hub to the VSAT. Satellite transponder access is FDMA

With demand assignment, frequency agility is required for transmitting VSATs only.

4.6.3.4 *FDMA–MCPC inbound/TDM outbound*

The number of modulators at the VSATs can be reduced to one, as illustrated in Figure 4.18, by time division multiplexing the traffic on the K inbound carriers

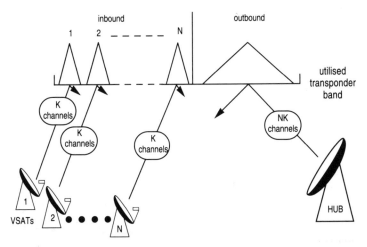

Figure 4.18 Star shaped network with multiplexed duplex connections conveyed on two carriers: one from the VSAT to the hub station, and one from the hub to the VSAT. Satellite transponder access is FDMA

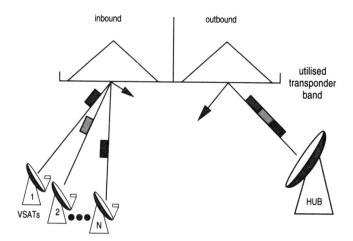

inbound   outbound

utilised
transponder
band

VSATs

HUB

Figure 4.19 Star shaped network with TDMA

from every VSAT to the hub station onto a single MCPC carrier. As the number of multiplexed connections on the inbound link may vary with time, the VSAT modulator must be at variable rate. Also, as the transmission rate is higher, the VSAT transmitter must be more powerful. The hub station needs to be equipped with N demodulators only.

With demand assignment, frequency agility is required for transmitting VSATs only.

4.6.3.5   TDMA inbound/TDM outbound

The VSAT may now access the satellite transponder in a TDMA mode, every VSAT transmitting its carrier burst in sequence, at the same bandwidth and the same frequency, as illustrated in Figure 4.19. Denoting by T_B the duration of a carrier burst and by T_F the duration of a frame, any VSAT transmits with a duty cycle T_B/T_F.

The capacity of a radio frequency link from a VSAT is equal to the number of transmitted bits per unit of time. In a TDMA scheme, if a VSAT is to benefit from the same radio frequency link capacity as with FDMA, then it has to transmit at a higher bit rate. Indeed, with FDMA, the radio frequency link capacity is equal to the continuous transmitted bit rate. With TDMA, the radio frequency link capacity of the VSAT is given by the number of bits transmitted per frame duration.

As can be seen from Figure 4.20, where R_{TDMA} and R_{FDMA} are the transmitted bit rates for respectively TDMA and FDMA, T_B is the burst duration and T_F the TDMA frame duration, the number of bits transmitted per frame duration is equal to $R_{TDMA}T_B$ for TDMA, while it is equal to $R_{FDMA}T_F$ for FDMA. Equating those two expressions leads to:

$$R_{TDMA} = R_{FDMA}\frac{T_F}{T_B}$$

(4.16)

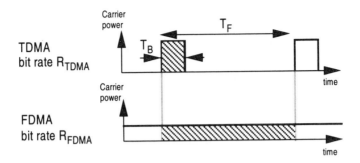

Figure 4.20 Comparison of bit rate and carrier power for FDMA and TDMA

Clearly the transmission rate is higher by a factor equal to the inverse of the duty cycle. If one neglects guard time between bursts, the inverse of the duty cycle is equal to the number of VSATs in the network. Therefore, for a given capacity, a large number of VSATs entails a high bit rate transmission.

It will be shown in Chapter 5 that the power of the carrier is proportional to the bit rate. Therefore, TDMA places a larger demand for power than FDMA from the VSAT transmitters.

Consider, for instance, a VSAT network with $N = 50$ VSATs, each with a radio frequency link capacity of 64 kb/s. With FDMA, all VSATs transmit at $R_{FDMA} = 64$ kb/s, and the satellite transponder bandwidth has to support 50×64 kb/s $= 3.2$ Mb/s. With TDMA, the same bandwidth would be used but now all VSATs would be requested to transmit at a rate of 3.2 Mb/s, thus with an increased demand for power by a factor of 50, or 17 dB, which is beyond the capability of cheap VSATs. Therefore, it would be necessary to reduce the capacity of every VSAT.

The following scheme, which is a hybrid one combining FDMA and TDMA, brings some flexibility to a cost effective design.

4.6.3.6   *FDMA–TDMA inbound/FDMA–MCPC outbound*

In order to lower the requirement on the VSAT transmitter power by reducing the transmitted bit rate, an elegant solution is to organise VSATs in groups, with L VSATs per group, a group sharing the same frequency band and accessing the satellite transponder in a TDMA mode. The different groups use different frequency bands: this is a *combined FDMA–TDMA* scheme, as illustrated in Figure 4.21. With this approach, given the number N of VSATs in the network and capacity per VSAT, the transmitted bit rate on the carrier burst, and hence the required carrier power, is divided by G, the number of groups.

For instance, in the previous example, by splitting the 50 VSATs in 5 groups of 10 VSATs the transmitted bit rate reduces from 3.2 Mb/s for pure TDMA to 640 kb/s, and the increase in power demand compared to pure FDMA is only 10 dB higher instead of 17 dB.

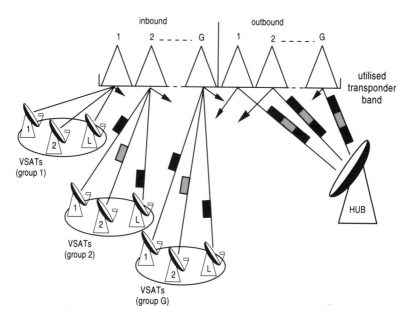

Figure 4.21 Star shaped network using a combined FDMA–TDMA inbound scheme, and an FDMA–MCPC outbound scheme

It may be convenient to consider time division multiplexing of all connections from hub to VSATs of the same group on one MCPC outbound carrier. The MCPC outbound carriers for the different groups then access the transponder in an FDMA mode. This reduces the bit rate transmitted by the hub, and hence its transmitter power, and offers the network manager the opportunity of implementing groups of VSATs as independent networks, sharing a common hub.

4.6.3.7   CDMA

Figure 4.22 illustrates the variety of schemes that can be considered in connection with full CDMA access, or a combination of CDMA and FDMA for the inbound and the outbound links. CDMA access can also be combined with SCPC or MCPC by grouping inbound connections.

With CDMA, carriers are assigned spreading pseudo-random codes instead of frequencies because all carriers use the same centre frequency. Hence frequency agility is no longer needed for demand assignment, eliminating the problem caused by frequency instability and phase noise encountered by SCPC/FDMA carriers that require precise frequency assignments. The major drawback to CDMA is its low throughput [MAR93, section 4.6.5, p. 162] which can be accepted only if it is balanced by the advantages gained from rejection of interference caused by other systems sharing the same frequency bands and polarisation.

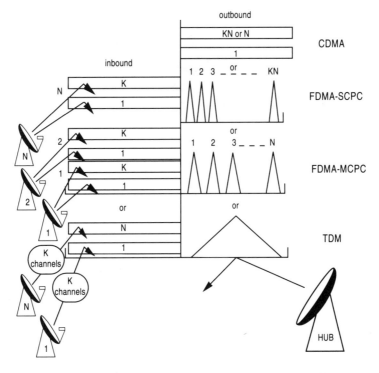

Figure 4.22 Star shaped VSAT network using CDMA or a combination of CDMA and FDMA

4.6.4 Fixed assignment versus demand assignment

Demand assignment has been presented in Chapter 1, section 1.5 as an appealing option within VSAT networks. It has been shown in the context of meshed networks (see section 4.6.2) that demand assignment permits implementing the desired connectivity between VSATs by setting up temporary links, with reduced VSAT equipment compared to fixed assignment allowing permanent links.

Therefore, it is interesting to discuss the impact of demand assignment, compared to fixed assignment in a general sense.

4.6.4.1 *Fixed assignment with FDMA (FA–FDMA)*

The network comprises N VSATs, each possibly transmitting K carriers at bit rate R_c. So, we have $L = KN$ carriers, and every carrier is allocated a given subband of the satellite transponder bandwidth. This subband is used by a VSAT when active (carrier 'on'), and remains unused when the VSAT has no traffic to convey (carrier 'off'). Should this happen, the capacity corresponding to the subband allocated to the VSAT is lost to the network. Figure 4.23 illustrates how fixed assignment works for FDMA in the case where $K = 1$.

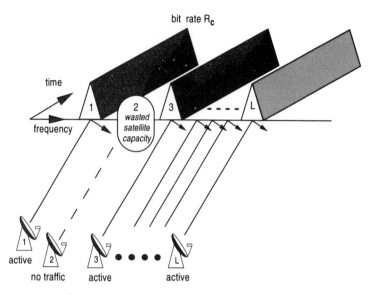

Figure 4.23   Fixed assignment FDMA ($K=1$)

Fixed assignment has the advantage of simplicity, and provides no blocking nor waiting time for setting up a carrier. However, the total network capacity of the VSAT network, which is equal to LR_c, is poorly utilised if the traffic demand is highly variable.

Call blocking may occur at a user terminal attached to a given VSAT, should several of these terminals wish to establish connections simultaneously with other terminals in the network, and the number of requested connections exceed the capacity of the VSAT.

For instance, consider the VSAT network of Figure 4.15, with $N_t = 8$ user terminals per VSAT, $N = 50$ VSATs and $K = 3$ transmitted 64 kb/s SCPC carriers per VSAT. The transponder bandwidth used by the inbound links is split into $L = KN = 3 \times 50 = 150$ subbands. Every subband is allocated to one 64 kb/s carrier. Therefore, the network capacity is $LR_c = 150 \times 64\,\text{kb/s} = 9.6\,\text{Mb/s}$. Assuming user terminals to generate traffic with intensity $A_t = 0.1$ erlang, then the traffic intensity offered to the $K = 3$ channels VSAT capacity is $A_{VSAT} = N_t A_t = 0.8$ erlang. The probability of call blocking, as given by formula (4.5), is equal to:

$$E_3(0.8) = 3.9 \times 10^{-2} = 3.9\%$$

4.6.4.2   Demand assignment with FDMA (DA–FDMA)

The network again comprises N VSATs, each possibly transmitting K carriers, and sharing a pool of L frequency subbands. But now $L < KN$. These subbands are used by the active VSATs. Figure 4.24 illustrates how demand assignment works

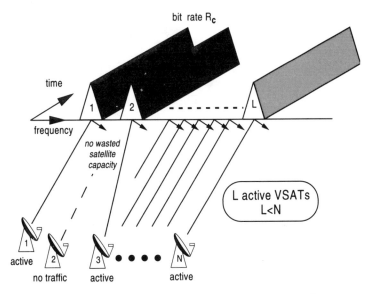

Figure 4.24 Demand assignment FDMA ($K = 1$)

for FDMA in the case where $K = 1$. Should the number of carriers exceed the number that can be supported by the allocated satellite transponder bandwidth, then there is blocking at the VSAT level: at the time of the call, no new carrier can be set up.

For example, again consider the VSAT network of Figure 4.15, with $N_t = 8$ terminals per VSAT, $K = 3$ transmitted 64 kb/s SCPC carriers per VSAT. Assuming terminals to generate traffic with intensity $A_t = 0.1$ erlang, then the traffic intensity offered to the $K = 3$ channels VSAT capacity is $A_{VSAT} = N_t A_t = 0.8$ erlang. Assuming again $N = 50$ VSATs, then the traffic intensity offered to the L subbands is $A = NA_{VSAT} = 50 \times 0.8 = 40$ erlang. The blocking probability for setting up a carrier can be maintained at a low level by having a sufficiently large pool of subbands. For instance, using (4.5), and taking $L = 60$ subbands, the carrier set-up blocking probability is $E_{60}(40) = 0.07\%$.

A call is blocked either because a terminal cannot access any one of the K channels, or because the VSAT cannot access any one of the L subbands. The probability for a call to be blocked is given by:

$$P_{blocked} = E_3(0.8) + E_{60}(40) - E_3(0.8)E_{60}(40)$$

$$= 3.9 \times 10^{-2} + 0.07 \times 10^{-2} - (3.9 \times 10^{-2}) \times (0.07 \times 10^{-2})$$

$$= 3.97\%$$

The capacity of the network is now $LR_c = 60 \times 64$ kb/s $= 3.84$ Mb/s. Therefore, with a negligible increase in call blocking probability, demand assignment offers a potential saving of $100(1 - L/KN)\% = 100(1 - 3.84$ Mb/s$/9.6$ Mb/s$) = 60\%$ of the used satellite transponder bandwidth.

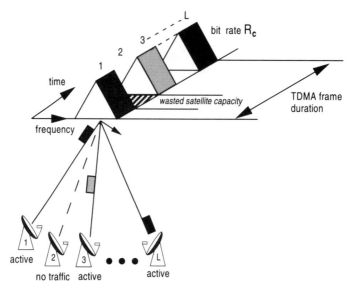

Figure 4.25 Fixed assignment TDMA

4.6.4.3 *Fixed assignment with TDMA (FA–TDMA)*

Figure 4.25 illustrates fixed assignment in connection with TDMA operation. Every VSAT transmits a carrier burst within a dedicated time slot. The number of time slots is equal to the number L of VSATs. Position and duration of bursts are fixed, therefore the capacity of every VSAT is constant whatever the traffic demand.

If R_c is the transmitted bit rate, then the total network capacity is R_c, and the capacity allocated to a VSAT is R_c/L. Should a VSAT have no traffic to transmit, then the slot remains unoccupied, and the corresponding capacity is lost for the network.

Fixed assignment has the advantage of simplicity, and provides no blocking nor waiting time for setting up a carrier. However, the total network capacity of the VSAT network (transponder bandwidth allocated to the network) is poorly utilised if the traffic demand is highly variable.

Blocking may occur at the user terminal. The blocking probability can be calculated following derivations similar to that of the example of fixed assignment with FDMA. One can assume that the carrier burst is split into K subbursts, each corresponding to one channel available to the attached user terminals.

Considering as previously $L = 50, K = 3, N_t = 8$ and $A_t = 0.1$ erlang, the blocking probability for a call is $E_3(0.8) = 3.9 \times 10^{-2} = 3.9\%$, as previously.

It is worth recalling that in order to achieve comparable capacities per VSAT, the value of R_c must be increased by a factor L relative to the FDMA scheme (see equation (4.16)).

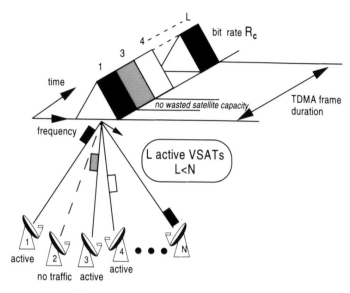

Figure 4.26 Demand assignment TDMA

4.6.4.4 *Demand assignment with TDMA (DA–TDMA)*

Figure 4.26 shows how the L time slots of the frame are now shared by N VSATs, with $N > L$. Any VSAT wishing to set up a link (carrier 'on' from carrier 'off' state) can access any unoccupied time slot on the frame, or should it already be active, it can increase its capacity by increasing the duration of its burst, and then support a larger number of connections.

This requires a change in the burst time plan, and is performed under the control of the network management system (NMS) at the hub station.

As the traffic demand from all VSATs may exceed the offered capacity R_c, blocking of link set-up may occur, as a result of the TDMA frame being filled with carrier bursts.

For example, the network capacity $KL = 150$ channels considered in the above FA–TDMA scheme is now available as a pool to all user terminals, whose total traffic intensity is $A = N \times 8 \times 0.1 = 0.8 N$ erlang, where N is the number of VSATs in the network. For a comparison with FA–TDMA, N can be selected so as to achieve a 4% blocking probability for a call. This means solving $E_{150}(A) = 4\%$, which corresponds to $A = 144$ erlang. Therefore $N = 144/0.8 = 180$, which indicates that the number of VSATs in the network can be increased by the factor $180/50 = 3.6$.

4.6.4.5 *Demand assignment multiple access (DAMA) procedure*

With demand assignment, a VSAT receives a call demand from one of the user terminals attached to it. This is indicated in Figure 4.27 as 'call arrival time'. This call may concern:

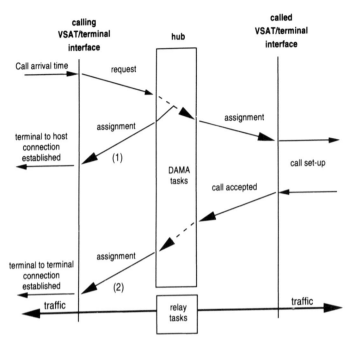

Figure 4.27 Demand assignment multiple access (DAMA) procedure: (1) terminal to host computer connection set-up, (2) terminal to terminal connection set-up

(1) an application at the host computer attached to the hub, or

(2) a user terminal attached to some other VSAT in the network.

As illustrated in Figure 4.27, the VSAT then sends a request to the hub by means of a specific inbound signalling channel, and the hub allocates the requested capacity, if available, to the corresponding VSAT by means of response messages transmitted on an outbound signalling channel, and conveying capacity assignments (carrier frequency, time slot or code).

If the connection is between a user terminal and the host computer, the delay of the incoming response is a two-hop delay, plus processing time. If the connection is between two user terminals, the hub transmits an assignment to the destination terminal and waits for a call acceptance message. This avoids assigning capacity to the incoming call before making sure the call can be accepted by the destination terminal. Then the hub assigns capacity to the calling VSAT. The connection set-up then takes a four-hop delay plus processing time. Once the connection is established from terminal to terminal, the hub acts as a relay.

Demand assignment requires that some network capacity be dedicated to request and response signalling. Moreover, some reservation delay is to be expected as a result of the time for signalling to be routed and processed.

When the access of VSATs to the signalling channel is organised according to a fixed assignment scheme, the need for limiting the capacity of the signalling channel to a reasonable fraction of the total network capacity will limit the number of VSATs in the network. In order to avoid such a limitation and offer the possibility of easy addition of new VSATs to the network, a random access scheme to the signalling channel is often preferred.

4.6.4.6 Demand assignment limitations

From the above, it can be seen that demand assignment allows more VSATs to share the satellite resource, or for a given network size allows a reduction in the utilised satellite bandwidth. Also to be considered is the penalty for signalling capacity. However, it can be kept small enough.

Of greater concern is the time delay: a message must wait at the VSAT before being transmitted while requests are forwarded and channel allocation is completed. As a result of propagation time and processing time, the delay may be as high as one to two seconds.

This is not compatible with efficient transmission of short messages which constitute bursty traffic, should a connection be set up every time a message comes in.

To illustrate the problem, consider Figure 4.28 where the connection set up delay is compared to the transmission time of a message (also called 'service time'). Table 4.1 indicates that the typical message length for bursty traffic is a few

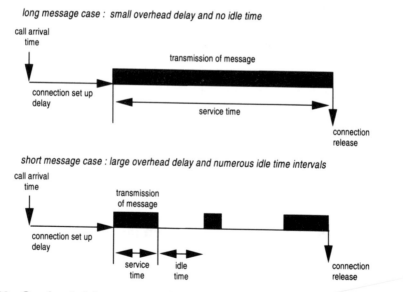

Figure 4.28 Overhead delay associated with connection set-up and idle time during connection for the case of a long message and that of a short one (bursty traffic)

hundred bytes. Considering, for example, a message length of 200 bytes, the service time at a rate of 64 kb/s is 25 ms. If one considers a connection set-up delay of 1.5 s, then the overhead delay before transmission of the message is as large as 60 times the message transmission time.

Moreover, with bursty traffic the temporary connection established on demand for message transfer is poorly utilised as a result of idle times. The larger the burstiness, the worse the problem. Indeed, considering an information transmission rate R_b and a message length L, the service time is $\tau = L/R_b$. Denote by $\langle IAT \rangle$ the average interarrival time, and recall the expression for burstiness from (4.8):

$$BU = R_b \frac{\langle IAT \rangle}{L} \qquad (4.17)$$

Then, from the definition of channel utilisation (section 4.2.7):

$$\text{Channel utilisation} = \frac{\text{service time}}{\text{service time} + \text{idle time}}$$

$$= \frac{\tau}{\langle IAT \rangle}$$

$$= \frac{L/R_b}{\langle IAT \rangle}$$

$$= \frac{1}{BU}$$

For instance, considering a value $BU = 10\,000$, the channel utilisation is $10^{-4} = 0.01\%$. With such a low utilisation of the channel, the advantage gained in terms of capacity reduction from demand assignment is lost.

Bursty traffic is routed most efficiently if:

—VSATs transmit at once whenever they get traffic from the user terminals. This forbids any connection set-up delay;

—the capacity derived from the utilisation of transponder bandwidth is shared between all VSATs to allow statistical multiplexing of the bursts.

These conditions are satisfied in random time division multiple access schemes, often named ALOHA.

4.6.5 Random time division multiple access

4.6.5.1 Principle

Random time division multiple access, also named ALOHA, has been introduced in section 4.6.1. There are two modes to it : unslotted ALOHA, and slotted

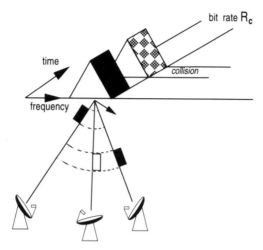

Figure 4.29   Principle of S-ALOHA

ALOHA (S-ALOHA) [HA86, pp. 362–378] [SCH77, Chapter 13]. With unslotted ALOHA, VSATs can transmit at any time, which means that they are not synchronised. With S-ALOHA, VSATs transmit in time slots, which means that they are synchronised but not coordinated in the sense that while transmitting in a given time slot, they ignore whether other VSATs are transmitting or not in the same time slot.

Figure 4.29 illustrates the principle of S-ALOHA. Every carrier is transmitted in the form of a burst with duration equal to that of a time slot. Every carrier burst conveys a packet of data. Synchronisation between VSATs is derived from the signals transmitted by the hub station and received on the outbound link.

Transmission of a packet is initiated by a message being generated by a user terminal attached to the VSAT. The length of the message may not coincide with the length of a packet. If too small, it must be completed by dummy bits. If too large, it must be conveyed over several packets. As users are not coordinated, messages as well as carrier bursts are generated at random.

The transmission efficiency of the S-ALOHA protocol is measured by the normalised throughput S, expressed as the number of packets successfully transmitted per packet length. The average rate of delivered bits $\langle R_c \rangle$, assuming packet length L bits, packet duration τ and transmitted bit rate $R_c = L/\tau$, is:

$$\langle R_c \rangle = S\frac{L}{\tau} = SR_c \tag{4.19}$$

It can be shown [TAN89 p. 22] [HA86, p. 362] that S depends on the average offered traffic, G (in packets per packet length), consisting of newly generated and retransmitted packets, and the number, N, of VSATs in the network:

$$S = G\left(1 - \frac{G}{N}\right)^{N-1} \tag{4.20}$$

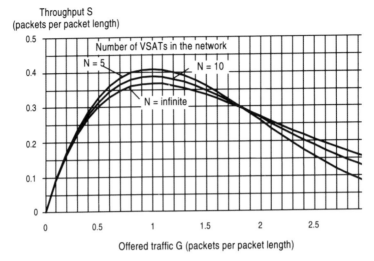

Figure 4.30 Throughput S versus offered traffic G for VSAT inbound link using S-ALOHA multiple access scheme

As N becomes infinite, the above expression becomes:

$$S = Ge^{-G} \tag{4.21}$$

These equations are represented in Figure 4.30.

From Figure 4.30, one can see that the normalised throughput converges rapidly to the infinite population case and that the maximum normalised throughput for an infinite number of VSATs is equal to $1/e = 0.368$, or 37%, which is poor.

With unslotted ALOHA, expression (4.21) is replaced by:

$$S = Ge^{-2G} \tag{4.22}$$

which leads to similar curves, but with an even lower maximum normalised throughput $S_{max} = 1/2e = 0.184 = 18\%$.

The low throughput of ALOHA schemes could be expected as VSATs may transmit at random in any time slot. The advantage of S-ALOHA is a higher throughput than unslotted ALOHA, and, as will be demonstrated in section 4.6.6, a lower average delay at low throughput (i.e. with highly bursty traffic), than that achieved with demand assignment FDMA or TDMA.

For a given transmission bit rate R_c(b/s), the number, N, of VSATs that can be installed in the network given the packet generation rate, λ(s^{-1}), per VSAT and the length, L, of a packet relates to the normalised throughput, S:

$$N = S\frac{R_c}{\lambda L} \tag{4.23}$$

The above expression is displayed in Figure 4.31 for various packet lengths and practical values of $S = 0.1$, and $R_c = 64$ kb/s.

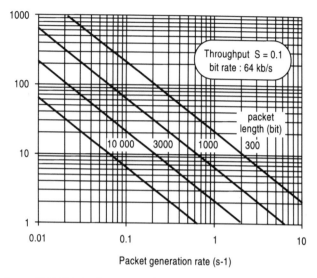

Figure 4.31 Number of VSATs in the network as a function of the packet generation rate for various packet lengths (in bits). A practical throughput $S = 0.1$ is considered, and a transmission bit rate $R_c = 64$ kb/s

4.6.5.2 Random access limitations

Instability: Packets suffering collisions must be retransmitted. VSATs awaiting retransmission are said to be 'backlogged'. The retransmitted packets generate an excess of traffic which adds to the newly generated packets. Therefore the offered traffic G exceeds the successful traffic, or throughput S. Referring to Figure 4.30, should the value of G exceed one, which corresponds to the maximum value of S, then as the throughput S decreases for any increase of G, collisions become even more frequent and the network evolves towards a situation of high G and low S where VSATs constantly retransmit the same packets.

Retransmission procedures are [BOE93]:

(1) Fixed retransmission probability: the backlogged terminal retransmits a packet with a fixed probability during each slot. This is simple but may be unstable [JEN80].

(2) Adaptive strategies: the feedback channel is observed and the retransmission probability is adapted according to the history of the channel.

(3) Heuristic retransmission: the retransmission probability is adjusted according to the number of retransmission attempts that have already been performed for the current packet.

In order to avoid instability, the design of a VSAT network should consider a practical throughput S of 5 to 15%. Figure 4.31, which considers $S = 0.1$ can be used to determine the appropriate number of VSATs.

Long message case: Successful transmission of long messages generated by the user terminals entails successful transmission of several consecutive packets. It can be shown that the throughput of an S-ALOHA scheme when long messages made of many consecutive packets must be retransmitted is given by [RAY84]:

$$S = \frac{Ge^{-G}}{1 + G^2} \tag{4.24}$$

The above formula indicates that the throughput compared to the case of single packet S-ALOHA, as given by expression (4.21), is lowered by a factor $(1 + G^2)$ and has a maximum value of 0.137 at $G = 0.414$.

   In order to overcome the above limitations, variants of the random time division multiple access schemes have been proposed and will now be presented.

4.6.5.3   Selective reject ALOHA

Messages are transmitted asynchronously as in unslotted ALOHA, but are partitioned into a finite number of short packets, each with its own acquisition preamble and header [RAY87] [RAY88]. The protocol exploits the fact that on an unslotted channel, most collisions are partial, so that uncollided portions of messages encountering conflict can be recovered by the receiver, and only the packets actually encountering conflict are retransmitted, in a way similar to the selective repeat protocol of Figure 4.6c. The maximum throughput has been shown to approach 0.368, irrespective of the message length distribution. In practice, the need for acquisition preamble and header in each packet limits the maximum useful throughput to the range of 0.2–0.3. The critical need for low packet overhead requires burst modems with short acquisition time.

4.6.5.4   Reservation/random TDMA

The inbound packets are transmitted in the time slots of a TDMA frame. This TDMA frame is shared by a group of VSATs in the network, as illustrated in Figure 4.21. The S-ALOHA protocol is used to convey messages short enough to be sent in a single packet. When an arriving message exceeds the slot size, a request for the necessary number of dedicated slots is made via S-ALOHA, while the remaining packets are queued at the VSAT waiting for a time slot assignment message from the hub for the remaining part of the message. All participating VSATs receiving the outbound carrier from the hub are informed of the specific slots reserved for the requesting VSAT and will refrain from

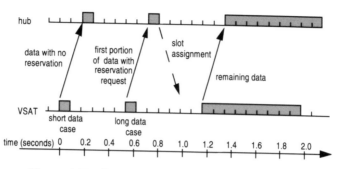

Figure 4.32 Reservation/random TDMA [FUJ86]

transmitting packets in these slots. The requesting VSAT now can transmit the remaining part of the message without collision.

The maximum number of time slots in a TDMA frame to be used for the reservation is limited to a predetermined value so that a certain number of slots are always kept free for random use. This precludes the occurrence of a lockout situation.

The requests for reservation can be either transmitted within a specific time slot divided into smaller slots, to be used on a contention basis with the ALOHA technique, within dedicated time slots [ROB73] [MOU92], or piggy-backed on the first packet of the message [FUJ86]. Figure 4.32 illustrates the latter case.

The arrival of a long message triggers the emission of a reservation message. Should the transmission of reservation messages be achieved on a contention basis using the S-ALOHA protocol, a high degree of correlation exists between the reservation traffic fluctuations and the data traffic load. This correlation may generate instability, as data traffic may display high instantaneous fluctuations. To alleviate the correlation, it has been proposed that a reservation message conveys reservation requests for more than one data message [ZEI91]. Not only does this reduce the load of the reservation channel, but the reservation protocol becomes unconditionally stable by simply indicating in the reservation message the number of data slots required for all the data messages waiting in the buffer of the VSAT.

4.6.5.5 *Random access with notification*

Satellite transponder capacity is divided into two portions: a signalling channel shared in fixed TDMA mode by multiple VSATs, each with a dedicated signalling slot, and a data channel to carry all data packets [CHI88] [CHI89]. Signalling and data channels are divided into fixed duration frames, and each frame is divided into a number of time slots. Frames and slots are numbered and synchronised among all VSATs via centralised control from the hub.

For the data channel, two types of slots are considered: random access and reservation. New packets use random access slots while collided and errored packets will be allotted reserved slots. Over the separate TDMA signalling channel, every VSAT notifies the hub of the number of packets transmitted by the VSAT in a particular frame. If a packet experiences a collision, the hub recognises that the number of packets successfully received is less than the number that were reported as sent. The hub then allots the required number of slots to the VSAT in some future frame for contention-free retransmission in the reservation mode.

4.6.6   Delay analysis

Delay means the time it takes for a message to be transferred from the user terminal to the host computer. The delay is a random variable, and it is common to characterise it by its average value. However, for many applications it is not sufficient to design a VSAT network on the basis of the average delay. Quantities such as the 95th percentile of delay are of equal or greater significance.

For delay analysis, assumptions are as follows. Every user generates messages according to a Poisson process (see Appendix 1). Messages are stored at the VSAT in a buffer of infinite capacity. A message is transmitted by the VSAT to the hub in the form of packets with length L.

4.6.6.1   Delay components

The message delay components for the inbound link and the outbound link are as follows:

(1) Delays at the VSAT:

—polling and/or queuing at the VSAT-to-user terminal interface;

—VSAT processing delay (formatting of packets);

—protocol induced delays: with random access, this is the waiting time for retransmission if a collision has been encountered; with demand assignment, this is the waiting time for allocation;

—time to send messages, called service time, which depends on the message size and the bit rate.

(2) Satellite link transmission delay: this corresponds to the propagation time of the carrier, and is given in Figure 2.22.

(3) Delays at the hub:

—polling and/or queuing at the hub-to-host interface;

—hub processing delay (formatting of packets);

—time to send messages, called service time, which depends on the message size and the bit rate.

The results presented below deal with packet delay, defined as the elapsed time between the instant when a packet is being transmitted by a source VSAT to the instant when the last byte of the packet is received by the destination hub. Processing time at the VSAT and at the hub is neglected.

It should be remarked that the considered packet delay is induced by the multiple access protocol only. It does not include delays possibly induced by retransmissions initiated by the data link control protocol. However, if the bit error rate on the satellite link is low enough, and selection of an appropriate data link control protocol has been made, retransmissions due to errors are rare. It can then be considered that the packet delay induced by the multiple access protocol is representative of the actual delay.

4.6.6.2 Comparison of FDMA, TDMA and S-ALOHA

Expressions for the average packet delay are as follows:

FDMA [HA86, p. 357] [TAN89, p. 636]:

(1) For exponentially distributed packet length:

$$T_{FDMA} = T_p + \frac{1}{\dfrac{R_{FDMA}}{L} - \lambda} \quad \text{(s)} \tag{4.25}$$

(2) For constant packet length:

$$T_{FDMA} = T_p + \frac{2 - \dfrac{\lambda}{\dfrac{R_{FDMA}}{L}}}{2\left(\dfrac{R_{FDMA}}{L} - \lambda\right)} \quad \text{(s)} \tag{4.26}$$

where T_{FDMA} is the average delay (s) for FDMA, T_p is the satellite link propagation delay (about 0.25 s), R_{FDMA} the carrier transmission rate (b/s), L the length of a packet (bits) and λ the average generation rate of packets (s^{-1}) per carrier.

If the network has N VSATs and every VSAT transmits K carriers, then the total network capacity is $R = NKR_{FDMA}$. Therefore, $R_{FDMA} = R/NK$.

TDMA [HA86, p. 361]:

(1) for exponentially distributed packet length:

$$T_{TDMA} = T_p + \frac{1}{\dfrac{R_{TDMA}}{NL} - \lambda} - \frac{T_F}{2} + \frac{T_F}{N} \quad \text{(s)} \tag{4.27}$$

(2) for constant packet length:

$$T_{TDMA} = T_p + \frac{2 - \dfrac{\lambda}{\dfrac{R_{TDMA}}{NL}}}{2\left(\dfrac{R_{TDMA}}{NL} - \lambda\right)} - \frac{T_F}{2} + \frac{T_F}{N} \quad (s) \qquad (4.28)$$

where T_{TDMA} is the average delay (s) for TDMA, T_p is the satellite link propagation delay (about 0.25 s), T_F the frame duration (s), L the length of a packet (bits), λ the average generation rate of packets (s^{-1}) per VSAT, R_{TDMA} the carrier transmission rate (b/s) and N the number of VSATs sharing the network capacity $R = R_{TDMA}$.

If the network has N VSATs, every VSAT transmits at a rate R_{TDMA}, equal to the network capacity R, and its time share of the capacity is R_{TDMA}/N.

Demand assignment FDMA and TDMA: Assuming there is no call blocking, the requested connection is set up after an initial delay equal to the round trip propagation delay equal to $2T_p$ (about 0.5 s). Once the connection is set up, the above formulas are applicable.

S-ALOHA [HA86, p. 369] [MOU92]:   Assuming the number of VSATs to be large enough (typically larger than 10) for equation (4.21) to hold:

$$T_{S\text{-}ALOHA} = T_p + \frac{3\tau}{2} + (e^G - 1)\left(2T_p + (k+1)\frac{\tau}{2} + \frac{\tau}{2}\right) \qquad (4.29)$$

where $T_{S\text{-}ALOHA}$ is the average delay (s) for S-ALOHA, T_p the satellite link propagation delay (about 0.25 s), τ the packet duration (s): $\tau = L/R_{TDMA}$, where R_{TDMA} is the carrier transmission rate (b/s), G the offered traffic load (packet per time slot), and k the maximum retransmission interval (time slots).

Figure 4.33 displays packet delay curves for a constant length packet traffic, with packet length $L = 1000$ bits, and packet generation rate $\lambda = 0.1\,s^{-1}$ as a function of the number of VSATs in the network. Three multiple access protocols are compared: FDMA with delay given by (4.26), TDMA with delay given by (4.28), and S-ALOHA with delay given by (4.29). For FDMA, it is assumed that a VSAT transmits one carrier only ($K = 1$). For S-ALOHA, three values of the maximum retransmission interval have been selected: $k = 10$, 50, 100 slots. The network capacity is constant and equal to $R = 100\,kb/s$. Thus, $R_{FDMA} = 100/KN = 100/N\,(kb/s)$, and $R_{TDMA} = R = 100\,kb/s$. The TDMA frame duration T_F varies with the number N of VSATs: $T_F = N \times L/R_{TDMA} = N \times 10^{-3}$ seconds.

Clearly, S-ALOHA delivers a shorter delay than FDMA and TDMA. This demonstrates its higher ability to convey bursty traffic. This has been confirmed by simulations [LIA92].

The normalised throughput S is given by:

$$S = \frac{N\lambda L}{R} \qquad (4.30)$$

where N is the number of VSATs, and R the network capacity. With the selected values, $S = 10^{-3}N$. A safe design with the S-ALOHA protocol would be to limit the number of stations below, say, 150, in order to maintain the throughput value below 15%.

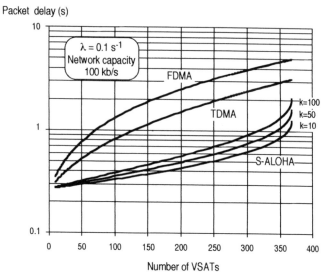

Figure 4.33 Packet delay for a constant length packet traffic, with packet length $L = 1000$ bits, and packet generation rate $\lambda = 0.1\,\mathrm{s}^{-1}$ per VSAT as a function of the number of VSATs in the network. The network capacity is $R = 100\,\mathrm{kb/s}$. With S-ALOHA, k represents the maximum retransmission interval in time slots

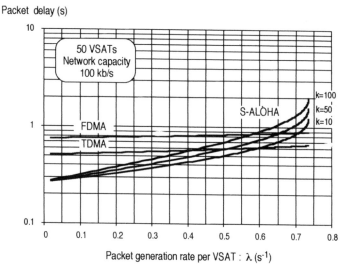

Figure 4.34 Average packet delay for a constant length packet traffic as a function of the packet generation rate λ per VSAT. Packet length is $L = 1000$ bits. The network comprises 50 VSATs. The network capacity is $R = 100\,\mathrm{kb/s}$. With S-ALOHA, k represents the maximum retransmission interval in time slots

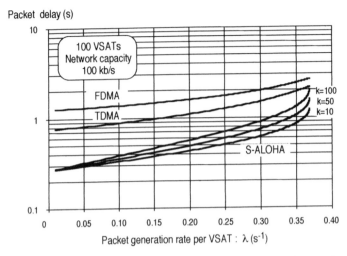

Figure 4.35 Average packet delay for a constant length packet traffic as a function of the packet generation rate λ per VSAT. Packet length is $L = 1000$ bits. The network comprises 100 VSATs. The network capacity is $R = 100$ kb/s. With S-ALOHA, k represents the maximum retransmission interval in time slots.

It is also of interest to view the average packet delay when the size of the network is given. Equations (4.26), (4.28) and (4.29) are used again for two network sizes: 50 VSATs (Figure 4.34) and 100 VSATs (Figure 4.35). In both figures the total network capacity is the same as with Figure 4.33, i.e. $R = 100$ kb/s.

4.6.6.3   Comparison of other protocols

Figure 4.36 presents a comparison of the delay performance of selective reject ALOHA with that of unslotted ALOHA, S-ALOHA and DAMA/TDMA [RAY88]. The latter corresponds to the variant of the reservation/random TDMA protocol where the requests for reservation are transmitted within a specific time slot divided into smaller slots, to be used on a contention basis using S-ALOHA [ROB73] [MOU92].

The results presented in Figure 4.36 have been obtained from computer simulations, considering a channel data rate of 56 kb/s, a message generation rate per VSAT of $0.07\,\mathrm{s}^{-1}$, a message length distribution adjusted to a truncated exponential law, with a maximum length of 256 bytes (2048 bits). Selective reject ALOHA outperforms all its competitors at low throughput (small size networks). As throughput increases (larger size networks), reservation/random TDMA displays a lower delay.

Figure 4.37 presents results obtained for the other variant of reservation/random TDMA protocol, where a reservation message is piggy-backed on the first packet of the message [FUJ86]. A mixture of random access traffic (short data) and

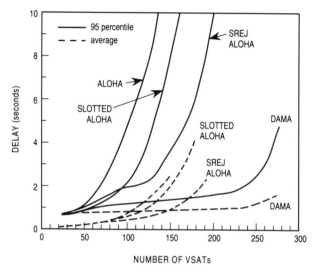

Figure 4.36 Average and peak delay versus number of VSATs for candidate multi-access protocols. (Reproduced from [RAY88] by permission of the Institute of Electrical and Electronics Engineers, Inc., © 1988 IEEE)

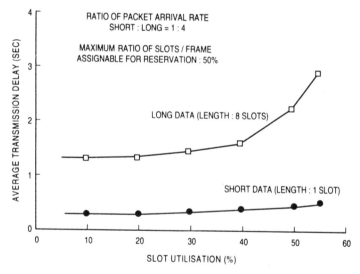

Figure 4.37 Delay versus throughput for mixed mode transmission. (Reproduced from [FUJ86] by permission of the Institute of Electrical and Electronics Engineers, Inc., © 1986 IEEE)

reservation traffic (long data) is transmitted. In the case shown here, the delay for the short data is minimised in spite of the coexistence of relatively large amounts of long data even where total slot utilisation is as high as 40%.

Figure 4.38 presents results obtained for the random access with noti-fication (RAN) protocol [CHI88] [CHI89]. It can be seen that the RAN

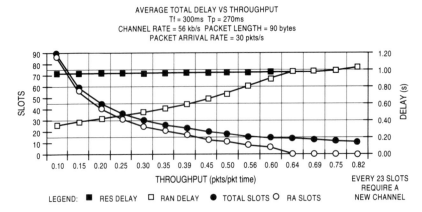

Figure 4.38 Performance of the Random Access with notification (RAN) protocol. (Reproduced from [CHI88] by permission of the Institute of Electrical and Electronics Engineers, Inc., © 1988 IEEE)

protocol has a better average delay than reservation TDMA, up to approximately 64% throughput, at which point the number of random access slots reaches zero, and the protocol transitions to a pure demand assignment TDMA scheme.

4.6.7 Conclusion

The above discussion provides guidelines for the selection of the multiple access technique appropriate to a given traffic type. It should be kept in mind that the selection of such a technique should maintain the implementation cost at a minimum, compatible with the required performance.

The combined FDMA–TDMA scheme of Figure 4.21 offers a convenient trade-off between the utilised transponder bandwidth and the power requirement for VSAT transmitters. It also allows a flexible implementation of techniques, from demand assignment to pure random access.

—For stream traffic, demand assigned TDMA offers a high throughput at the expense of an acceptable delay.

—For bursty traffic, a dedicated connection would be poorly used, and random time division multiple access is more efficient. S-ALOHA offers low delay at high burstiness, and can be combined with reservation techniques in order to avoid the inherent limitations of S-ALOHA at low burstiness.

These techniques are used in most present VSAT networks that support a mix of both interactive and batch applications.

CDMA is well suited for low data rate applications, to meet interference restrictions, and for applications that require spectrum spreading for frequency coordination (for instance C-band VSAT networks).

Other protocols, not discussed here because they are too complex for a cost effective implementation in VSAT networks, have been proposed. The extensive overview provided in [RAY87] makes a highly recommendable reference.

4.7   NETWORK DESIGN

4.7.1   Principles

Network design aims at translating the customer's requirements in terms of information transport performance and network availability into network configuration, space segment requirements, backup hardware elements and channel configurations.

Most of the customer's requirements have been discussed in Chapter 3, section 3.3. It has been mentioned how important it is to start with reliable information on traffic characteristics and to get a clear perspective of the user's performance goals in terms of link availability, response time, etc.

Fortunately, most often VSAT networks replace existing leased line data networks, so that the designer can start with a comprehensive network diagram and user locations. He also needs to know the detailed characteristics of the host processing system: type, communications hardware interface, number of circuits, access protocols and applications software.

Traffic data are more difficult to obtain. Ideally, what is needed is, by type of application and by user location: the number of busy hour transactions, the average size of transaction in and out, the time of day for file transfer and for interactive traffic, etc.

The customer will be asked to specify his performance objectives : average and percentile (for example 95%) delay for interactive applications, and for some applications, special requirements such as delay jitter. His requirements on availability will condition radio frequency link margins and terrestrial backup connections. It will also determine backup hardware, maintenance policy and backup services such as shared or portable hub, portable VSATs and alternative remote network management systems.

From these inputs, the designer will define the entire hardware configuration at the hub and at remote VSATs with respect to number of ports, bit rate, protocols and other parameters. The design should make due allowance for possible expansion during the years following the initial installation and operation.

The satellite space segment will be defined in detail. The following information should be included: type of carriers and transmitted bit rate per carrier and number of inbound carriers per outbound (TDM) carrier, total number of inbound and outbound carriers, and multiple access control with respect to contention, reservation or a combination thereof.

The network design is then used to determine the cost of the network, with possible feedback to the customer's requirements.

4.7.2 Guidelines for preliminary dimensioning

The design of a network can be supported by guidelines as an outcome from simulations performed on a prevalent network architecture [ZEI91a]. An example is now presented which deals with the typical star network shown in Figure 4.39, for interactive enquiry–response applications (see Table 1.2).

Every VSAT has a number of cluster control units (CCUs) connected to it via local or intermediate range terrestrial links (access lines). The bit rate on these access lines is referred to as R_{cv}. Every CCU, in turn, has a number of computer terminals (or display terminals) attached to it. The host computer is connected to the hub station via a front end processor (FEP).

Due to the low volume and bursty nature of VSAT enquiry traffic, inbound transponder access protocol is slotted-ALOHA. The outbound link operates in a TDM (Time Division Multiplex) mode. The bit rates of the inbound and outbound links are referred to as R_{vh} and R_{hv} respectively. The hub station is connected to the FEP through a high capacity terrestrial link. The bit rate of this link is referred to as R_{hf}.

Polling emulation, as described in section 4.5.3, is used in the two terrestrial links. The data link control over the satellite link is kept simple by assuming that every enquiry message represents an independent packet over the inbound link. Error-free links are assumed, and retransmissions take place due to collisions in

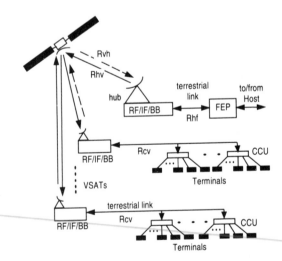

Figure 4.39 Network configuration. RF: Radio frequency; FEP: Front End Processor; IF: Intermediate Frequency; CCU: Cluster Control Unit; BB: Baseband; R_{cv}: bit rate in CCU–VSAT link; R_{vh}: bit rate in VSAT–hub link; R_{hf}: bit rate in hub–FEP link; R_{hv}: bit rate in hub–VSAT link

the random access. When collisions occur, the packets are scheduled to be retransmitted after a random delay chosen from a uniform distribution. Positive acknowledgement of successful enquiry messages is required since VSATs cannot reliably receive their own transmissions. This implies a minimum time-out delay of approximately 0.5 to 0.6 s before a retransmission can be initiated. A flow control process is also required in the inbound channel to prevent congestion or deadlock situations. This is achieved by a sliding window technique with a window size of seven messages.

The enquiry message length is 20 bytes and the response message length is 256 bytes. However, since the messages are transacted by a data link control layer, there is additional overhead associated with transmission of every message. Further, one considers a single response for every inputted enquiry. The generation of enquiry messages at the CCUs is approximated by Poisson processes.

The network dimensioning refers to the process of determining the minimum bit rate required for R_{cv}, R_{vh}, R_{hf} and R_{hv}, given the network configuration and the average response time required. The network configuration includes the number of VSATs (N), the number of CCUs per VSAT (M), and the average message generation rate at every CCU (λ msg/h).

The CCU–VSAT and the hub–FEP links can each be viewed as a bus conveying altogether enquiry and response messages. The length of the enquiry message is 26 bytes (20 data bytes and 6 bytes for addressing) and the length of the response message is 256 bytes. The offered traffic on these links is:

$$\text{CCU–VSAT link:} \quad T_{cv}(b/s) = M\lambda \frac{(26 + 256) \times 8}{3600} \tag{4.31}$$

$$\text{hub–FEP link:} \quad T_{hf}(b/s) = M\lambda N \frac{(26 + 256) \times 8}{3600} \tag{4.32}$$

The VSAT–hub inbound link conveys enquiry messages only, with a length equal to 32 bytes (26 as delivered by the CCU and 6 bytes for addressing). The hub–VSAT outbound link conveys response messages only, with a length equal to 262 bytes (256 data bytes and 6 bytes for addressing). The offered traffic on each of these links is:

$$\text{VSAT–hub link:} \quad T_{vh}(b/s) = M\lambda N \frac{32 \times 8}{3600} \tag{4.33}$$

$$\text{hub–VSAT link:} \quad T_{hv}(b/s) = M\lambda N \frac{262 \times 8}{3600} \tag{4.34}$$

The bit rate of every link will now be normalised to the offered traffic of this link. The network is first dimensioned for a given traffic configuration. The resulting dimensioning parameters are then used for other traffic configurations. This allows one to determine the limits of validity of the parameters and to derive dimensioning guidelines.

It is considered that an acceptable response time should be of the order of 2 to 3 seconds. In the following, the dimensioning procedure will be demonstrated

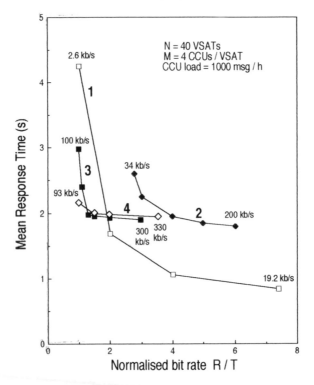

Figure 4.40 Mean response time versus bit rate of network links normalised to traffic volume in the link. Curve 1: Effect of R_{cv} ($R_{vh} = 200$, $R_{hf} = 300$, $R_{hv} = 300$ kb/s). Curve 2: Effect of R_{vh} ($R_{cv} = 4.8$, $R_{hf} = 300$, $R_{hv} = 300$ kb/s). Curve 3: Effect of R_{hf} ($R_{cv} = 4.8$, $R_{vh} = 46$, $R_{hv} = 300$ kb/s). Curve 4: Effect of R_{hf} ($R_{cv} = 4.8$, $R_{vh} = 46$, $R_{hf} = 130$ kb/s)

considering a specified mean response time of 2.5 seconds. Of course, the procedure is valid for other values of the mean response time.

Figure 4.40 displays the effect of the bit rate of any link on the mean delay. These results are obtained by simulation.

Curve 1 illustrates the influence of the CCU–VSAT link bit rate, R_{cv}, on the network response time. All other links are overdimensioned to reduce their influence. This curve shows that the CCU–VSAT link bit rate has a strong influence on the overall response time. Note that for $R_{cv} = 4800\,\text{bit/s} \approx 2T_{cv}$ (normalised bit rate = 2), the overall average delay is less than 2 seconds, and it is needless to exceed that normalised bit rate.

Curve 2 illustrates the effect of the VSAT–hub link bit rate, R_{vh}, on the network response time setting $R_{cv} = 4800\,\text{bit/s} \approx 2T_{cv}$, and for overdimensioned hub–FEP and hub–VSAT links. It can be noticed that when the inbound link bit rate exceeds four times the offered traffic in this link ($R_{vh} = 46\,\text{kb/s} \approx 4T_{vh}$), the effect of the bit rate of this channel on the global average network response time becomes negligible. One of the reasons for this good behaviour of the inbound link operated in slotted-ALOHA is that small fixed length messages are considered for the enquiry traffic. If variable length and/or long enquiry messages were considered, the inbound link would have had more effect on the global average network response time.

Curve 3 illustrates the effect of the hub–FEP link bit rate, R_{hf}, on the network response time. This curve is obtained for $R_{cv} = 4800\,b/s \simeq 2T_{cv}$ and for $R_{vh} = 46\,kb/s \approx 4T_{vh}$. The hub-VSAT link is kept overdimensioned. The curve shows that when the capacity of the hub–FEP link exceeds 1.3 times the offered traffic in this link, the link has no more influence on the delay. From the selection of the above values, this link can be dimensioned with $R_{hf} = 130\,kb/s \approx 1.3T_{hf}$.

Curve 4 illustrates the effect of the hub–VSAT link bit rate, R_{hv}, on the network response time. This curve is obtained for a hub–FEP link bit rate of $130\,kb/s \approx 1.3T_{hf}$. The bit rates of the CCU–VSAT link and the VSAT–hub link are the same as for curve 3. The curve indicates that the hub–VSAT link displays a very small effect on the network response time. Hence, the bit rate of the outbound link can be dimensioned to be equal to the offered traffic $T_{hv}(R_{hv} = T_{hv} = 93\,kb/s)$.

In conclusion, the normalised bit rates required in the four links are:

2 in the CC–VSAT link
4 in the VSAT–hub link
1.3 in the hub–FEP link
1 in the hub–VSAT link.

The resulting mean response time with this dimensioning is 2.16 s, which fulfils the 2.5 s objective. If a smaller delay is desired, only the bit rate of the CCU–VSAT link needs to be increased since increasing the bit rates of the other links does not really improve the overall delay.

The above dimensioning procedure is only intended to provide good approximations of the required bit rates. Giving the exact values is of little interest as, in practice, one has to use standardised bit rates (4.8, 9.6, 19.2, 56, 64 kb/s, etc.). For example, if the above dimensioning procedure leads to a required bit rate in the CCU–VSAT link of 6 kb/s, the actual value to be implemented is the next higher standard bit rate which is 9.6 kb/s. Therefore, the actual response time should be smaller than the response time obtained by simulation.

The sensitivity of this dimensioning has been explored and is reported in [ZEI91a]. Different CCU loads and numbers of VSATs have been considered. It has been shown that, to avoid too large a delay for a practical range of parameter variations, one is led to consider a minimum network dimension, that is, $R_{cv} = 4800\,bit/s$, $R_{vh} = 19.2\,kb/s$, $R_{hf} = 56\,kb/s$, and $R_{hv} = 56\,kb/s$.

In conclusion, given any network configuration (M, N, λ) and a delay requirement of 2.5 seconds, the network is dimensioned as follows:

(1) Compute the offered T_{vh}, T_{hf}, T_{hv} according to formulas (4.31) to (4.34).

(2) Compute the preliminary bit rate required in each link: $R_{cv} = 2T_{cv}$, $R_{vh} = 4T_{vh}$, $R_{hf} = 1.3T_{hf}$, and $R_{hv} = T_{hv}$.

(3) Select, for each bit rate, the next higher standard bit rate.

(4) If any of the resulting bit rates is smaller than the minimum bit rate (i.e. $R_{cv} < 4800\,\text{b/s}$, $R_{vh} < 19.2\,\text{kb/s}$, $R_{hf} < 56\,\text{kb/s}$, or $R_{hv} < 56\,\text{kb/s}$), select the minimum bit rate.

4.7.3   Example

This procedure will now be applied to an example.

Consider a network characterised by $N = 50$ VSATs, $M = 6$ CCU/VSAT, and CCU load $\lambda = 500\,\text{msg/h/CCU}$. This traffic configuration results in the following offered traffic in the four network links: $T_{cv} = 1880\,\text{b/s}$, $T_{vh} = 10.7\,\text{kb/s}$, $T_{hf} = 94\,\text{kb/s}$, and $T_{hv} = 87.4\,\text{kb/s}$. Using the dimensioning guidelines of the last section, we have:

(1) The bit rate of the CCU–VSAT link R_{cv} should be twice T_{cv} ($R_{cv} = 3760\,\text{b/s}$). However, this value is smaller than the minimum bit rate considered for this link which is $4800\,\text{b/s}$. Hence, select for this link in the network a bit rate of $4800\,\text{b/s}$.

(2) The bit rate of the inbound channel R_{vh} should be at least four times the offered traffic T_{vh} in this link. The required bit rate is then equal to $42.8\,\text{kb/s}$. Hence, select a bit rate of $56\,\text{kb/s}$ which is the next higher standard bit rate.

(3) The bit rate of the hub–FEP link should be 1.3 times T_{hf}. This gives $R_{hf} = 122.2\,\text{kb/s}$. The next higher standard bit rate is $128\,\text{kb/s}$.

(4) In the hub–VSAT link, the required bit rate is equal to the offered traffic T_{hv} ($87.4\,\text{kb/s}$). Use the next higher standard bit rate, i.e. $128\,\text{kb/s}$.

Running the simulation with these values gives an average response time of $1.9\,\text{s}$, which fulfils the requirement of an average response time less than $2.5\,\text{s}$.

4.8   CONCLUSION

This chapter has described the organisation of a VSAT network, both its physical configuration and its protocol configuration.

It has been shown that protocols that perform well on terrestrial networks may cause poor channel efficiency when used on satellite links, as a consequence of higher delay. Therefore, it is important to perform protocol conversion at the interface between the hub station and the host computer or the remote terminals and the VSATs.

A critical aspect for achieving good throughput is the error recovery technique which must be compatible with the satellite channel delay and error characteristics. Error recovery may be carried out at the link level where the VSATs and the hub station assume the retransmission. An alternative is to recover errors at the transport level where retransmission of packets in error is performed by the remote terminals or the host computer.

Implementation of link level error recovery requires large transmitting and receiving buffers at VSATs and hub that can handle the number of packets equivalent to at least one round trip time. Doing it at the transport level avoids such buffers but introduces additional packet delay, as the time-out interval is longer.

It has been shown that the inherent delay (0.25 s for one hop) and the typical bit error rate encountered on satellite links (10^{-6}–10^{-7} for 99.9% of time) are no severe impediment to providing the user terminal with an acceptable service quality.

One must be aware of how important it is to select a multiple access protocol according to the user application. For stream traffic, demand assignment FDMA or TDMA is a good choice. FDMA systems, especially SCPC systems, allow operating at a low transmission rate, hence with low transmit power requirement. This is in favour of low VSAT equipment cost. Unfortunately, multicarrier operation of the satellite transponder leads to unavoidable generation of inter-modulation products, as will be discussed in the next chapter, unless operating the transponder at reduced power. But this may offset the advantage gained from FDMA. TDMA systems are known to exhibit the best efficiency and the highest flexibility, but at the expense of a high transmission rate. The best choice lies in between, combining FDMA and TDMA: for instance, VSATs can be organised in groups operating at different frequencies, with VSATs of a group sharing a given frequency band and accessing the satellite transponder in TDMA (Figure 4.21).

For bursty traffic, it has been shown that a combination of random TDMA and reservation techniques allows one to cope with a high range of burstiness. For instance, a combined scheme of S-ALOHA for transport of small data packets (high burstiness) and slot reservation for transport of long messages (low bursti-ness) gives today's VSAT networks the required flexibility for supporting a mix of both interactive and batch applications to the best user satisfaction.

Today, CDMA has not yet found widespread distribution in the non-military world, apart from some low data rate applications, to meet interference restric-tions, and for applications that require spectrum spreading for frequency coor-dination (for instance C-band VSAT networks).

5 RADIO FREQUENCY LINK ANALYSIS

The previous chapter has addressed the networking techniques which allow connection set-up and reliable transfer of information from one user terminal to another. This chapter will address information transfer at the *physical level*.

Figure 5.1 reproduces an excerpt from Figure 4.5 in order to put into perspective the respective topics of Chapter 4 and the present one. Chapter 4 dealt with the peer layers of the hub and VSAT interface within the VSAT network at data link control and satellite channel access control levels. The present chapter focuses on the physical layer, which involves forward error correction (FEC), modulation and coding.

Indeed, the satellite channel conveys information by means of modulated radio frequency carriers which are relayed by the satellite transponder and then received by the destination station. Noise contaminates the received carriers. Therefore, the retrieved baseband signals are also contaminated: analogue signals are noisy, and data may contain erroneous bits.

Basically, it is not feasible to provide error-free transmission at the physical layer level. The only hope is to limit the bit error rate (BER) to an acceptable level constrained by cost considerations. It is the job of the upper layers, and especially the data link layer, to ensure error-free transmission by means of automatic repeat request protocols. The job is easier when the physical layer already provides 'clean' information, thanks to a low enough bit error rate. As the BER decreases, the performance of the channel improves, as illustrated in Figure 4.7.

This chapter aims at providing the means to calculate the quality of the information contents delivered to the data link control layer. The quality of digital information is measured by the bit error rate (BER), which is the ratio of the number of bits received in error to the total number of received bits. The bit error rate depends on the type of modulation and coding performed, and on the carrier to noise power spectral density ratio at the input of the receiver. This ratio, C/N_0, can be considered as a quality measure of the radio frequency link.

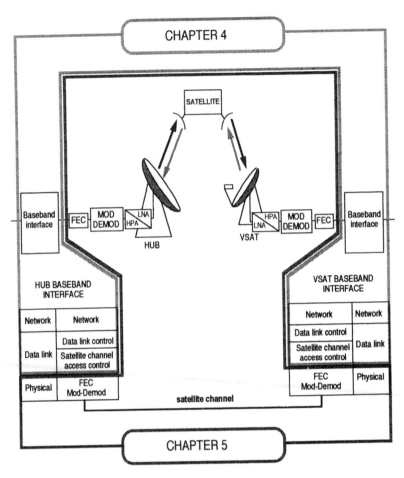

Figure 5.1 Topics covered in Chapters 4 and 5, respectively. FEC: forward error correction, MOD: modulation; DEMOD: demodulation; HPA: high power amplification, LNA: low noise amplification.

5.1 PRINCIPLES

The link analysis will be performed in the context of a star shaped network, as illustrated in Figure 5.2. The transmitting VSATs located within the coverage of the receiving antenna of the satellite generate N inbound carriers. These carriers are relayed by the satellite transponder to the hub station. The hub station communicates with the VSATs by means of a single outbound carrier which is modulated by a time division multiplex (TDM) stream of bits received by all VSATs within the coverage of the transmitting antenna of the satellite, thanks to the broadcasting capability of the satellite within its coverage area.

A carrier originating from a transmitting station and received by the satellite transponder at the uplink frequency, is amplified by the satellite transponder and

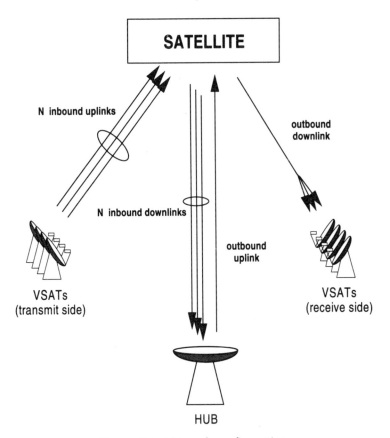

Figure 5.2 Network configuration.

frequency translated before being transmitted and received by the earth stations
tuned to the downlink frequency. This carrier is corrupted by noise with different
origins as discussed here below.

5.1.1 Thermal noise

Thermal noise is present on the uplink and the downlink and is produced by
natural sources. First, we have the radiation produced by radiating bodies and
captured by the receiving antennas. With satellite communications, the principal
sources of radiation are the Earth for the satellite antenna, and the sky for the earth
station antenna. Such a noise is called 'antenna noise'. Another source of thermal
noise is the noise generated by the receiver components.

5.1.2 Interference noise

Some noise interference is to be expected from systems sharing the same frequency bands, either satellite-based systems or terrestrial ones. Interference

introduces on the uplink, where the receiving satellite antenna is illuminated by carriers transmitted by earth stations belonging to an adjacent geostationary satellite system, or by terrestrial microwave relays. Interference also introduces on the downlink where the receiving earth station antenna captures carriers transmitted by adjacent satellites or terrestrial microwave relays. This interference acts as noise if the undesired carrier spectrum overlaps with that of the wanted carrier. The problem may be of special importance on the downlink as the small size of the VSAT station and its resulting large beamwidth makes it more sensitive to reception of off-axis carriers. This is why it is preferable that VSAT networks operate within exclusive bands (see section 1.5.4).

The above interference is generated by transmitters others than those operating within the considered VSAT network. Interference is also generated within the considered VSAT network. This is sometimes called 'self-interference'. For example, some VSAT networks incorporate earth stations operating on two orthogonal polarisations at the same frequency. It may also be that the satellite is a multibeam satellite, with stations transmitting on the same polarisation and frequency but in different beams. These techniques are referred to as 'frequency reuse' techniques and are used to increase the capacity of satellite systems without consuming more bandwidth [MAR93, p. 183]. However, the drawback is an increased level of interference due to imperfect cross polarisation isolation of antennas in the case of frequency reuse by orthogonal polarisation, and imperfect beam to beam isolation in the case of spatial frequency reuse.

5.1.3   Intermodulation noise

From Figure 5.2, one can see that the satellite transponder supports several carriers, either N if the inbound carriers and the outbound one are fed to separate transponders, or $N + 1$ if they share the same transponder, which is usually the case. With an access scheme like TDMA, where carriers are transmitted sequentially within a frame period (see Figure 4.19), only one of these carriers is amplified at a given instant by the transponder. However, with access schemes such as FDMA (see figures 4.15 to 4.18) or CDMA (see Figure 4.22), where carriers are continuously transmitted by the earth stations, the transponder amplifies several carriers simultaneously in a so-called 'multicarrier mode'. This is also the case when a hybrid access mode such as FDMA/TDMA is implemented (see Figure 4.21). The difference resides in the number of simultaneous carriers. This has two consequences : firstly, the output power of the satellite transponder is shared between the simultaneous carriers and this reduces by as many the power available to each carrier; secondly, the presence of simultaneous carriers in the non-linear amplifying device of the transponder causes the generation of intermodulation products in the form of signals at frequencies f_{IM} which are linear combinations of the P input frequencies [MAR93, p. 132]. Thus:

$$f_{IM} = m_1 f_1 + m_2 f_2 + \cdots + m_P f_P \quad \text{(Hz)} \tag{5.1}$$

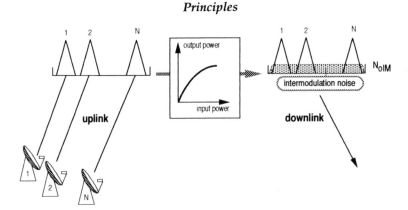

Figure 5.3 Multicarrier operation with FDMA access by N carriers.

where $m_1, m_2, \ldots, m_P$ are positive or negative integers. The quantity X, called the order of an intermodulation product, is defined as:

$$X = |m_1| + |m_2| + \cdots + |m_P| \tag{5.2}$$

As the centre frequency of the pass-band transponder is large compared with its bandwidth, only those odd-order intermodulation products with $\Sigma m_i = 1$ fall within the channel bandwidth. Moreover, the power of the intermodulation products decreases with the order of the product. Thus, in practice, only third-order products and, to a lesser extent, fifth-order products are significant. Inter-modulation products are transmitted on the downlink along with the wanted carrier, but no useful information can be extracted from them. They act as noise, as a fraction of the overall intermodulation product power falls into the bandwidth of the earth station receiver tuned to the wanted carrier. They can be modelled as white noise, with constant power spectral density N_{0IM} given by:

$$N_{0IM} = \frac{N_{IM}}{B_N} \quad \text{(W/Hz)} \tag{5.3}$$

where N_{IM} is the intermodulation power measured at the transponder output within the equivalent noise bandwidth B_N of the earth station receiver.

Figure 5.3 illustrates the above discussion and shows how intermodulation products can be accounted for in the form of an equivalent white noise with power spectral density equal to N_{0IM}.

5.1.4 Carrier power to noise power spectral density ratio

It should be clearly understood that these noise contributions are to be considered in relationship to the wanted carrier they corrupt. Therefore, it is useful to specify carrier power to noise power spectral density ratios at the point in the link where noise corrupts the carrier.

Figure 5.4 indicates at which point in the link a given quantity is relevant, and Table 5.1 specifies notations and definitions for the corresponding carrier power

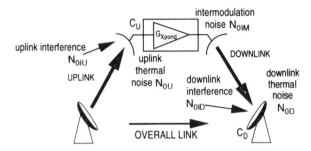

Figure 5.4 Point in the link where quantities used in Table 5-1 are relevant.

Table 5.1 Carrier power to noise power spectral density according to the considered noise contribution (see Figure 5.4 for point in the link where notation and definition applies)

Origin of noise	Notation for noise power spectral density (W/Hz)	Notation and definition for carrier power to noise power spectral density ratio
Uplink thermal noise	N_{0U}	$(C/N_0)_U = C_U/N_{0U}$
Downlink thermal noise	N_{0D}	$(C/N_0)_D = C_D/N_{0D}$
Uplink interference	N_{0iU}	$(C/N_{0i})_U = C_U/N_{0iU}$
Downlink interference	N_{0iD}	$(C/N_{0i})_D = C_D/N_{0iD}$
Intermodulation noise	N_{0IM}	$(C/N_0)_{IM} = G_{Xp0nd}G_U/N_{0IM}$

to noise power spectral density ratio. In Table 5.1 and Figure 5.4 G_{Xpond} is the power gain of the transponder for carrier power C_U at the transponder input.

5.1.5 Total noise

At the receiver input of the receiving station of Figure 5.4, the demodulated carrier power is C_D, and the power spectral density N_{0T} of the corrupting noise is that of the total noise contribution. This total contribution builds up from the following components.

—The uplink thermal noise and uplink interference noise retransmitted on the downlink by the satellite transponder. On their way from the satellite transponder input to the earth station receiver input, they are subject to power gains and losses which amount to a total gain G_{TE}. This power gain is the product of the transponder power gain G_{Xpond} and the gain G_D from transponder output to earth station receiver input (which in practice is much less than one in absolute value, so it should actually be considered as a loss):

$$G_{TE} = G_{Xpond} \times G_D \tag{5.4}$$

Therefore, the respective contributions of uplink thermal noise and uplink interference noise at the earth station receiver input are $G_{TE}N_{OU}$ and $G_{TE}N_{OiU}$. Note that G_{Xpond} has been defined as the transponder power gain for carrier power C_U at transponder input. The transponder being non-linear, the actual transponder gain depends on the power of the considered input signal. Hence, G_{Xpond} has different values for the noise and for the carrier. This is referred to as the 'capture effect'. However, for simplicity, this will not be considered here.

—The intermodulation noise generated at the transponder output and transmitted on the downlink. Hence, its contribution at the earth station receiver input is $G_D N_{0IM}$.

—The downlink thermal noise and downlink interference with respective contributions at the earth station receiver input N_{0D} and N_{0iD}.

The total noise power spectral density at the earth station receiver input is given by:

$$N_{OT} = \sum N_{0j} \quad (W/Hz) \tag{5.5}$$

where N_{0j} are the above individual contributions at the earth station receiver input.

The carrier power to noise power spectral density ratio at the earth station receiver input C_D/N_{OT} conditions the quality of the baseband signal delivered to the user terminal in terms of BER. This ratio relates to the overall link from station to station and will be denoted $(C/N_0)_T$ (T for total). $(C/N_0)_T$ can be calculated as follows:

$$\left(\frac{C}{N_0}\right)_T^{-1} = \frac{N_{OT}}{C_D} = \frac{\sum_j N_{0j}}{C_D}$$

$$= \frac{G_{TE}N_{0U} + G_{TE}N_{0iU}}{C_D} + \frac{G_D N_{0IM}}{C_D} + \frac{N_{0D}}{C_D} + \frac{N_{0iD}}{C_D} \quad (Hz^{-1}) \tag{5.6}$$

Consider that $C_D = G_{TE}C_U = G_{Xpond} \times G_D \times C_U$, equation (5.6) becomes:

$$\left(\frac{C}{N_0}\right)_T^{-1} = \frac{N_{0U}}{C_U} + \frac{N_{0iU}}{C_U} + \frac{N_{0IM}}{G_{Xpond}C_U} + \frac{N_{0D}}{C_D} + \frac{N_{0iD}}{C_D}$$

$$= \left(\frac{C}{N_0}\right)_U^{-1} + \left(\frac{C}{N_0}\right)_D^{-1} + \left(\frac{C}{N_0}\right)_{IM}^{-1} + \left(\frac{C}{N_{0i}}\right)_U^{-1} + \left(\frac{C}{N_{0i}}\right)_D^{-1} \quad (Hz^{-1}) \tag{5.7}$$

Note that introducing the actual transponder gain depending on signal power at transponder input instead of the same value G_{Xpond} for the noise and for the carrier would introduce a corrective term to the values of $(C/N_0)_U$ and $(C/N_{0i})_U$ in the above equation.

The following sections provide means for the determination of the terms implied in the calculation of $(C/N_0)_T$ according to equation (5.7). Sections 5.2 and 5.3 discuss the parameters involved in the calculation of uplink $(C/N_0)_U$ and downlink $(C/N_0)_D$. Section 5.4 discusses the intermodulation and the parameters involved in the calculation of $(C/N_0)_{IM}$. Section 5.5 is dedicated to interference analysis and means to calculate $(C/N_{0i})_U$ and $(C/N_{0i})_D$. Section 5.6 recapitulates the previous terms in expression (5.7) for the overall link $(C/N_0)_T$. Section 5.7 deals with bit error rate determination. Section 5.8 demonstrates how power and bandwidth can be exchanged through the use of forward error correction. Section 5.9 gives an example of calculation for VSAT networks.

5.2   UPLINK ANALYSIS

Figure 5.5 illustrates the geometry of the uplink. In order to calculate the value of $(C/N_0)_U$ in the worst case, the transmitting earth station is assumed to be located at the edge of the uplink coverage defined as the contour where the satellite receiving antenna has a constant gain defined relative to its maximum value at boresight, for instance $-3\,$dB, corresponding to a reduction by a factor two of the gain compared to its maximum. From Table 5.1, the ratio $(C/N_0)_U$ is defined as:

$$\left(\frac{C}{N_0}\right)_U = \frac{C_U}{N_{0U}} \quad (\text{Hz}) \tag{5.8}$$

where C_U is the power of the received carrier at the input to the satellite transponder.

N_{0U} is the noise power spectral density and relates to the uplink system noise temperature T_U, given by (5.32):

$$N_{0U} = kT_U \quad (\text{W}/\text{Hz}) \tag{5.9}$$

where k is the Boltzmann constant: $k = 1.38 \times 10^{-23}\,$J/K; $k(\text{dBJ}/\text{K}) = 10\log k = -228.6\,$dBJ/K.

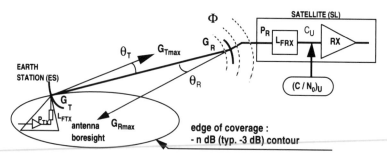

Figure 5.5   Geometry of the uplink P_{TX}:   transmitter output power; L_{FTX}:   feeder loss from transmitter to antenna; G_T: earth station antenna transmit gain in direction of satellite; θ_T: earth station antenna depointing angle; G_{Tmax}: earth station antenna transmit gain at boresight; Φ: power flux density at satellite antenna; G_R: satellite antenna receive gain at edge of coverage; θ_R: satellite antenna half beamwidth angle; P_R received power at antenna output ; L_{FRX}: feeder loss from satellite antenna to receiver input; C_U: carrier power at receiver input; RX: receiver.

$(C/N_0)_U$ can be expressed [MAR93, p. 65] as:

$$\left(\frac{C}{N_0}\right)_U = IBO_1\left(\frac{C}{N_0}\right)_{Usat} \quad (Hz)$$

$$\left(\frac{C}{N_0}\right)_U (dBHz) = IBO_1(dB) + \left(\frac{C}{N_0}\right)_{Usat} (dBHz) \quad (5.10)$$

where IBO_1 is the input back-off per carrier for the considered carrier, as defined in Appendix 6 by expression (A6.5), and $(C/N_0)_{Usat}$ is the value required to saturate the satellite transponder, and is given by:

$$\left(\frac{C}{N_0}\right)_{Usat} = \Phi_{sat}\frac{1}{G_1}\left(\frac{G}{T}\right)_{SL}\frac{1}{k} \quad (Hz)$$

$$\left(\frac{C}{N_0}\right)_{Usat} (dBHz) = \Phi_{sat}(dBW/m^2) - G_1(dBi) + \left(\frac{G}{T}\right)_{SL} (dBK^{-1}) - 10\log k\,(dBJ/K)$$

$$(5.11)$$

where Φ_{sat} is the power flux density required to saturate the satellite transponder (see section 5.2.1), $(G/T)_{SL}$ is the figure of merit of the satellite receiving equipment (see section 5.2.4), and G_1 is the gain of an ideal antenna with area equal to $1\,m^2$:

$$G_1 = \frac{4\pi}{\lambda^2} = 4\pi\left(\frac{f}{c}\right)^2$$

$$G_1(dBi) = 10\log 4\pi - 20\log \lambda = 10\log 4\pi + 20\log\left(\frac{f}{c}\right) \quad (5.12)$$

λ is the wavelength (m), f is the frequency (Hz), c is the speed of light: $c = 3 \times 10^8\,m/s; k$ is the Boltzmann constant: $k = 1.38 \times 10^{-23}\,J/K$; $k(dBJ/K) = 10\log k = -228.6\,dBJ/K$.

5.2.1 Power flux density at satellite distance

The power flux density Φ is defined in Appendix 5. Assume the satellite to be at distance R from a transmitting earth station, with effective isotropic radiated power $EIRP_{ES}$. Then the power flux density at satellite level is:

$$\Phi = \frac{EIRP_{ES}}{4\pi R^2} \quad (W/m^2)$$

$$\Phi(dBW/m^2) = EIRP_{ES}(dBW) - \log 4\pi R^2 \quad (5.13)$$

The power flux density can also be calculated from G_1 given by (5.12) and the uplink path loss L_U, discussed in section 5.2.3:

$$\Phi = \frac{EIRP_{ES}G_1}{L_U} \quad (W/m^2)$$

$$\Phi(dBW/m^2) = EIRP_{ES}(dBW) + G_1(dBi) - L_U(dB) \quad (5.14)$$

Satellite transponder input characteristics are given in terms of saturated power flux density Φ_{sat}, which designates the power flux density required to saturate the transponder. From expression (5.13), it can be seen that Φ is controlled by the transmitting earth station $EIRP_{ES}$ dedicated to the carrier. The power of that carrier at the transponder input determines IBO or IBO_1, as defined in Appendix 6 by equations (A6.2) and (A6.5) respectively. For example:

$$IBO_1 = \Phi_1 / \Phi_{sat}$$

$$IBO_1(dB) = \Phi_1(dBW/m^2) - \Phi_{sat}(dBW/m^2) \tag{5.15}$$

Should N stations be transmitting simultaneously, the powers of their individual carriers add at the transponder input, and the total flux density is given by the sum of their individual contributions, each calculated from (5.13) or (5.14):

$$\Phi_t = \Sigma\Phi_i \quad i = 1, 2, \ldots, N$$

The transponder total input back-off, defined in Appendix 6 by equation (A6.7), is then given by:

$$IBO_t = \frac{\Phi_t}{\Phi_{sat}}$$

$$IBO_t(dB) = \Phi_t(dBW/m^2) - \Phi_{sat}(dBW/m^2) \tag{5.16}$$

5.2.2   Effective isotropic radiated power of the earth station

From the definition given in Appendix 5, the effective isotropic radiated power of the earth station $EIRP_{ES}$ is expressed as:

$$EIRP_{ES} = P_T G_T \quad (W)$$

$$EIRP_{ES}(dBW) = P_T(dBW) + G_T(dBi) \tag{5.17}$$

where P_T is the power fed to the transmitting antenna, and G_T is the earth station antenna transmit gain in the pertinent direction.

Figure 5.6 is an enlargement of the transmitting earth station represented in Figure 5.5. The earth station transmitter TX with output power P_{TX} feeds power P_T to the antenna through a feeder with feeder loss L_{FTX}. The antenna displays a transmit gain G_{Tmax} at boresight, and a reduced transmit gain G_T in the direction of the satellite as a result of the transmit depointing off-axis angle θ_T. The transmitter output power P_{TX} is smaller than or equal to the transmitter output rated power P_{TXmax}, depending on the transmitter output back-off.

In order to calculate the actual gain G_T, one needs to know more about the antenna gain pattern. Appendix 4 defines the antenna gain pattern and its most

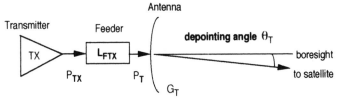

Figure 5.6   Transmitting earth station components.

important parameters, the maximum gain G_{max} and the half power beamwidth θ_{3dB}, which are expressed as:

$$G_{max} = \eta_a \left(\frac{\pi Df}{c} \right)^2$$

$$G_{max}(dBi) = 10 \log \left[\eta_a \left(\frac{\pi Df}{c} \right)^2 \right] \tag{5.18}$$

and

$$\theta_{3dB} = 70 \frac{c}{fD} \quad \text{(degrees)} \tag{5.19}$$

where:

η_a = antenna efficiency (typically 0.6)
D = antenna diameter (m)
f = frequency (Hz)
c = speed of light = 3×10^8 m/s

Figures 5.7 and 5.8 display values of these parameters for typical hub and VSAT antenna diameters. For the transmit gain and half power beamwidth values, one should consider 14 GHz and 6 GHz for the frequency values.

Should there be no feeder loss, the power fed to the antenna would be P_{TX}, and should the antenna be perfectly pointed, its transmit gain in the direction of the satellite would be equal to G_{Tmax}. Therefore, $EIRP_{ES}$ would be maximum and equal to:

$$EIRP_{ESmax} = P_{TX} G_{Tmax} \quad \text{(W)}$$

$$EIRP_{ESmax}(dBW) = P_{TX}(dBW) + G_{Tmax}(dBi) \tag{5.20}$$

Figure 5.9 displays the maximum EIRP values that can be achieved from a given combination of transmitter power and antenna diameter. This is to be considered as an upper bound, as provided by an ideal transmitting equipment perfectly pointed. The actual EIRP value depends on the magnitude of the losses, which will now be discussed.

ANTENNA GAIN (dBi) = 10 log η (π Df / c)^2 , η = 0.6, c= 3 x 10^8 m/s

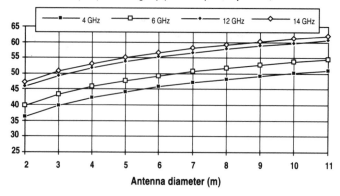

HALF POWER BEAMWIDTH = 70 c / Df (degrees)

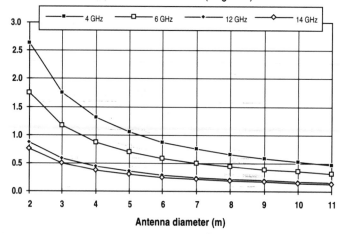

Figure 5.7 Antenna gain and half power beamwidth for typical hub station antenna diameter.

The power P_T fed to the antenna can be calculated from the known values of the transmitter output power P_{TX} and the feeder loss L_{FTX}:

$$P_T = \frac{P_{TX}}{L_{FTX}} \quad (W)$$

$$P_T(dBW) = P_{TX}(dBW) - L_{FTX}(dB) \tag{5.21}$$

Typically, L_{FTX} is of the order of 0.1 to 0.3 dB.

Similarly, G_T can be calculated from the maximum value of the earth station antenna gain G_{Tmax} and the transmission depointing loss L_T:

$$G_T = \frac{G_{Tmax}}{L_T}$$

$$G_T(dBi) = G_{Tmax}(dBi) - L_T(dB) \tag{5.22}$$

ANTENNA GAIN (dBi) = 10 log η (π Df / c)^2 , η = 0.6, c= 3 x 10^8 m/s

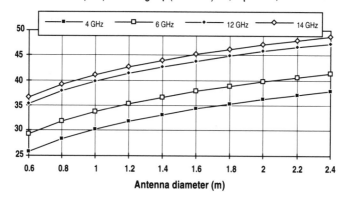

HALF POWER BEAMWIDTH = 70 c / Df (degrees)

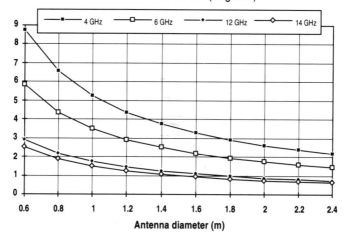

Figure 5.8 Antenna gain and half power beamwidth for typical VSAT antenna diameter.

where

$$G_{Tmax} = \eta_a \left(\frac{\pi Df}{c} \right)^2$$

$$G_{Tmax}(\text{dBi}) = 10 \log \left[\eta_a \left(\frac{\pi Df}{c} \right)^2 \right] \tag{5.23}$$

and

η_a = antenna efficiency (typically 0.6)
D = antenna diameter (m)
f = frequency (Hz)
c = speed of light = 3×10^8 m/s

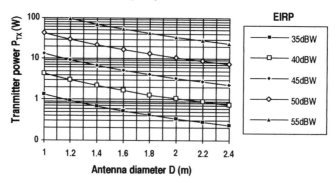

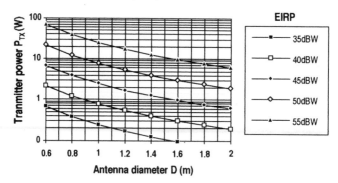

Figure 5.9 Maximum achievable EIRP$_{ESmax}$ at 6 GHz and 14 GHz for given antenna size and transmitter power.

The depointing loss L_T can be expressed in dB as a function of the depointing angle θ_T and the transmit half power beamwidth θ_{3dB} of the antenna:

$$L_T(dB) = 12\left(\frac{\theta_T}{\theta_{3dB}}\right)^2 \tag{5.24}$$

where θ_{3dB} is given by (5.19).

The depointing angle θ_T is not easy to determine. Should the earth station be equipped with a tracking antenna, as would a large hub station, say with antenna diameter larger than 5 m at Ku-band and 9 m at C-band, the depointing angle is given by the tracking accuracy of the tracking equipment, and is typically of the order of $0.2\theta_{3dB}$. Therefore, the depointing loss remains smaller than 0.5 dB. But small hub stations and VSATs are equipped with fixed mount antennas. The value of the maximum depointing angle θ_{Tmax} then depends on the pointing accuracy at installation, and the subsequent motion of the geostationary satellite. An upper limit to the depointing angle value can be estimated from an analysis of the pointing procedure and the satellite motion.

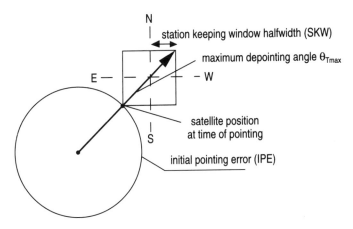

Figure 5.10 Geometry of pointing errors.

Figure 5.10 represents the underlying geometry. The figure displays off-axis angles as viewed from the antenna boresight, represented by the centre of the circle. As indicated in Chapter 3, section 3.2.3, at time of installation the antenna is pointed towards the satellite by a search for a maximum of received power from a satellite beacon or a downlink carrier. Once this is done, the antenna is left to itself, and one assumes that the mount is robust enough to ensure a constant pointing direction with time. The radius of the circle of Figure 5.10 represents the maximum initial pointing error (IPE). Now the satellite is maintained within a so-called station keeping window (see Chapter 2, section 2.3.9). It is assumed here that the station keeping window dimensions are equal in the north–south (NS) direction and in the east–west (EW) one, so that the station keeping window can be represented by the square in Figure 5.10 with half-width SKW. The worst situation occurs when the initial pointing is performed while the satellite is at an extreme position within its station keeping window, say the SE corner. Subsequently, the satellite may move to the opposite extreme position, say the NW corner. Therefore, the maximum depointing angle θ_{Tmax} is the sum of the initial depointing error IPE and the angle corresponding to the extreme positions of the satellite, that is the diagonal of the station keeping window.

$$\theta_{\text{Tmax}} = \text{IPE} + 2\sqrt{2}\,\text{SKW} \quad \text{(degrees)} \tag{5.25}$$

The initial pointing error IPE is typically equal to $0.2\theta_{3\text{dB}}$, where $\theta_{3\text{dB}}$ represents the half power beamwidth of the antenna at the frequency of the received beacon or carrier used for pointing, i.e. at *downlink* frequency. Denoting this frequency by f_D and combining (5.19), (5.24) and (5.25) results in:

$$L_{\text{Tmax}} = 12\left[0.2\frac{f_U}{f_D} + 2\sqrt{2}\,\text{SKW}\frac{Df_U}{210 \times 10^8}\right]^2 \quad \text{(dB)} \tag{5.26}$$

Figure 5.11 displays the maximum transmit depointing loss L_{Tmax} for a C-band and a Ku-band system, assuming $\text{SKW} = 0.025°$, and Figure 5.12 displays the same curves, assuming $\text{SKW} = 0.05°$.

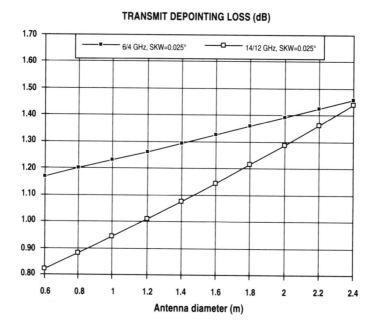

Figure 5.11 Transmit antenna gain depointing loss L_T for a C-band and a Ku-band system, assuming the half width of the station keeping window SKW = 0.025°.

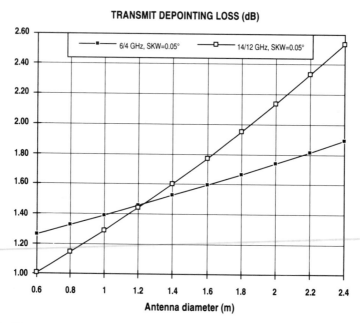

Figure 5.12 Transmit antenna gain depointing loss L_T for a C-band and a Ku-band system, assuming the half width of the station keeping window SKW = 0.05°.

Table 5.2 Maximum achievable EIRP, typical magnitude of losses, and actual EIRP for hub and VSATs: Ku-band (14 GHz)

	Large hub	Small hub	VSAT	
Antenna diameter D	10 m	3 m	1.8 m	1.2 m
Transmitter power P_{TX}	100 W	10 W	1 W	1 W
Maxmimum EIRP	81.1 dBW	61.6 dBW	46.2 dBW	42.7 dBW
Feeder loss L_{FTX}	0.2 ± 0.1 dB	0.2 ± 0.1 dB	0.2 ± 0.1 dB	0.2 ± 0.1 dB
Depointing loss L_T	0.5 ± 0.1 dB	2.4 ± 0.7 dB	1.6 ± 0.3 dB	1.2 ± 0.2 dB
Actual EIRP	80.4 ± 0.2 dBW	59 ± 0.8 dBW	44.4 ± 0.4 dBW	41.3 ± 0.3 dBW

Comparing the curves of Figures 5.11 and 5.12 provides some insight on the impact on depointing loss of the satellite station keeping window size.

Table 5.2 summarises the above results by providing typical values.

5.2.3 Uplink path loss

The uplink path loss, L_U, is the overall attenuation of the carrier power on its way from the earth station transmitting antenna to the satellite receiving antenna. It can be shown that this attenuation has two components, the free space loss, L_{FS}, defined in Appendix 5, and the atmospheric loss, L_A, so that the path loss can be expressed as [MAR93, Chapter 2]:

$$L_U = L_{FS} L_A$$
$$L_U(dB) = L_{FS}(dB) + L_A(dB) \tag{5.27}$$

The free space loss depends on the frequency f and on the distance R between the earth station and the satellite:

$$L_{FS} = \left(\frac{4\pi R f}{c}\right)^2 = \left(\frac{4\pi R_0 f}{c}\right)^2 \left(\frac{R}{R_0}\right)^2$$

$$L_{FS}(dB) = 10\log\left(\frac{4\pi R_0 f}{c}\right)^2 + 10\log\left(\frac{R}{R_0}\right)^2 \tag{5.28}$$

where c is the speed of light ($c = 3 \times 10^8$ m/s) and R_0 is the satellite height ($R_0 = 35\,786$ km for a geostationary satellite).

The ratio $(R/R_0)^2$ is a geometric factor which takes into account the position of the earth station relative to the sub-satellite point on the earth surface, and is expressed as:

$$\left(\frac{R}{R_0}\right)^2 = 1 + 0.42(1 - \cos l \cos L) \tag{5.29}$$

where l and L are respectively the difference in latitude and in longitude between the earth station and the sub-satellite point. Notice that the sub-satellite point of a geostationary satellite is on the equator, and hence its latitude is zero, so l identifies with the earth station latitude, while L should be taken as the actual difference in longitude between that of the earth station and that of the satellite meridian.

For a geostationary satellite, Figure 5.13 gives the variation in dB of the first term of equation (5.28), as a function of frequency, and Figure 5.14 gives the variation in dB of the second term of equation (5.28), as a function of the earth station location. The free space loss L_{FS} (in dB) is calculated by adding the values obtained from those two figures. The second term appears to be a small corrective term to the first one.

The attenuation of radio frequency carriers in the atmosphere, denoted by L_A, is due to the presence of gaseous components in the troposphere, water (rain, clouds, snow and ice) and the ionosphere. Water plays an important role especially at Ka-band as it has an absorption line at 22.3 GHz. Gaseous components and water in the form of vapour are constantly present in the atmosphere. Water is occasionally present in the form of rain and as such produces attenuation and cross-polarisation of the radio wave, i.e. transfer of part of energy transmitted in one polarisation to the orthogonal polarisation state.

It is convenient to consider power loss L_A as the result of two attenuation terms:

$$L_A = A_{AG} A_{RAIN}$$

$$L_A(dB) = A_{AG}(dB) + A_{RAIN}(dB) \tag{5.30}$$

where A_{AG} is the always present attenuation due to the atmosphere during 'clear sky' conditions (no rain) and A_{RAIN} is the additional and occasional attenuation due to rain.

The attenuation A_{AG} depends on frequency and elevation angle, and is higher at low elevation angles as a result of the increased path length of the radio wave in the atmosphere. One can consider that the attenuation A_{AG} is, for elevation angles greater than $10°$, negligible at C-band, less than 0.5 dB at Ku-band, and less than 1 dB at Ka-band.

The attenuation A_{RAIN} is to be considered in relationship to rainfall rate expressed in mm/hour. A_{RAIN} increases with rainfall rate and can reach high values when small percentages of time are considered. Rainfall rate depends on the climate and hence on the considered region of the world. Relevant techniques for the determination of rain attenuation A_{RAIN} for various time percentages are presented in ITU-R Reports 563, 564, 721 and 723. For planning purposes, the world has been divided into regions with similar climatic conditions. This classification is available from ITU-R Report 563.

As an example, Figure 5.15 shows the climatic regions that have been retained for Europe. The letters A to L correspond to increased values of rainfall rate which are exceeded for an annual time percentage of 0.01%. Figures 5.16 and 5.17 display the values of A_{RAIN} at C-band and Ku-band respectively for the climatic regions of

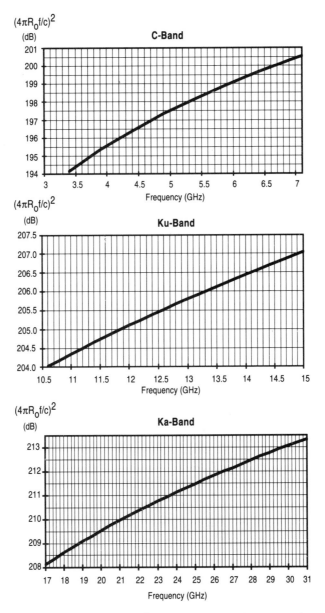

Figure 5.13 Variation in dB of $(4\pi R_0 f/c)^2$ as a function of frequency for a geostationary satelllite ($R_0 = 35\,786$ km).

Europe identified in Figure 5.15. One should notice that A_{RAIN} is sensitive to the elevation angle, as shown by the reduction of the attenuation as the elevation angle increases above $10°$. This results from the reduced path length of the radio wave through rain with increasing elevation angle.

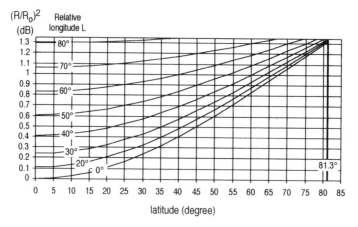

Figure 5.14 Variation in dB of $(R/R_0)^2$ as a function of the position of the earth station with respect to the sub-satellite point for a geostationary satellite.

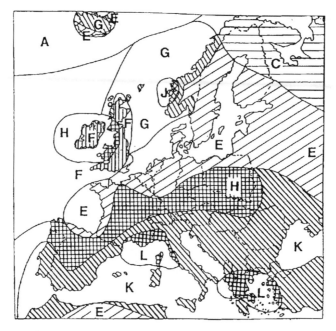

Figure 5.15 Climatic regions for Europe [MOR88].

Attenuation due to rain clouds and fog is usually small compared with that due to rain precipitation except for clouds and fog with a high water concentration. For an elevation angle $E = 20°$, it is negligible at C-band, typically 0.5 to 1.5 dB at Ku-band GHz, and 2 to 4 dB at Ka-band. This attenuation, however, is observed

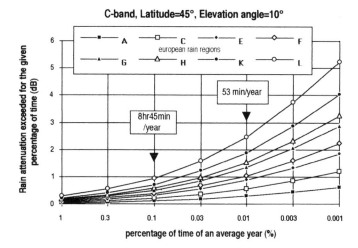

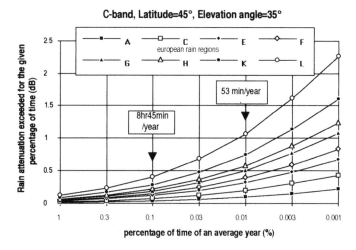

Figure 5.16 Values of A_{RAIN} at C-band for Europe.

for a greater time percentage. Attenuation due to ice clouds is smaller still. Dry snow has little effect. Although wet snowfalls can cause greater attenuation than the equivalent rainfall rate, this situation is very rare and has little effect on attenuation statistics. The degradation of antenna characteristics due to accumulation of snow and ice on the dish may be more significant than the effect of snow along the path.

Another effect of rain is depolarisation of the wave. As a consequence, where frequency reuse by orthogonal polarisation is used, a part of the carrier power transmitted in one polarisation is transferred to the orthogonal polarisation, causing cross-polarisation interference. This is discussed in section 5.5 in relation to interference.

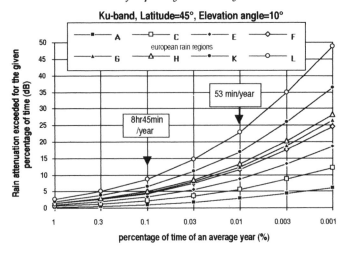

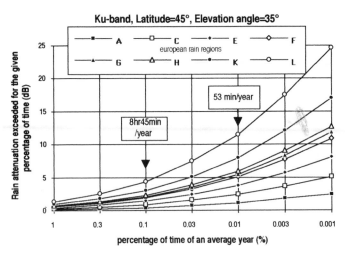

Figure 5.17   Values of A_{RAIN} at Ku-band for Europe.

5.2.4   Figure of merit of satellite receiving equipment

The figure of merit $(G/T)_{SL}$ of the satellite receiving equipment incorporates the composite gain from the antenna to the satellite receiver input, and the uplink system noise temperature. As a factor in the expression (5.10) or (5.11) for $(C/N_0)_U$, it indicates the capability of the satellite receiving equipment to build up a high value of $(C/N_0)_U$. Its expression is given by:

$$\left(\frac{G}{T}\right)_{SL} = \left(\frac{G_{Rmax}}{L_R L_{pol}}\right)_{SL} \left(\frac{1}{L_{FRX}}\right)_{SL} \left(\frac{1}{T_U}\right) \quad (K^{-1})$$

$$\left(\frac{G}{T}\right)_{SL} (dBK^{-1}) = G_{Rmax}(dBi) - L_R(dB) - L_{pol}(dB) - L_{FRX}(dB) - 10\log T_U \quad (5.31)$$

where G_{Rmax} is the satellite antenna receive gain at boresight, and L_R is the off-axis gain loss corresponding to reception from a station at edge of coverage. If the considered coverage contour corresponds to that of the half-power beamwidth, then $L_R = 2$ and $L_R(dB) = 3\,dB$. L_{pol} is the gain loss as a result of possible polarisation mismatch between the antenna and the received wave. Methods for evaluating this loss are given in [MAR93, pp. 26, 336]. A practical value is $L_{pol} = 0.1\,dB$. L_{FRX} is the feeder loss from the antenna to the receiver input, typically 1 dB. T_U is the uplink system noise temperature, and is given by:

$$T_U = \frac{T_A}{L_{FRX}} + T_F\left(1 - \frac{1}{L_{FRX}}\right) + T_R \quad (K) \tag{5.32}$$

where T_A is the satellite antenna noise temperature, T_F is the temperature of the feeder, and T_R is the satellite receiver effective input noise temperature.

Practical values are $T_A = 290\,K$, $T_F = 290\,K$, $L_{FRX} = 1\,dB$, $T_R = 500\,K$, hence $T_U = 790\,K$. The maximum satellite antenna gain G_{Rmax} depends on the coverage, typically 20 dBi for global coverage, 38 dBi for a narrow spot beam coverage. Therefore, $(G/T)_{SL}$ ranges from $-13\,dBK^{-1}$ for a global coverage to $+5\,dBK^{-1}$ for a spot beam coverage as indicated in Table 2.2.

5.3 DOWNLINK ANALYSIS

Figure 5.18 illustrates the geometry of the uplink. In order to calculate the worst case value of $(C/N_0)_D$, the transmitting earth station is assumed to be located at the edge of the downlink coverage defined as the contour where the satellite receiving antenna has a constant gain defined relative to its maximum value at boresight, for instance $-3\,dB$, corresponding to a reduction by a factor of two of the gain

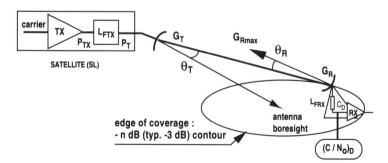

Figure 5.18   Geometry of the downlink. P_{TX}: carrier power at satellite transmitter output; L_{FTX}: feeder loss from satellite transmitter to antenna; P_T: carrier power fed to the satellite antenna; G_T: satellite antenna transmit gain in direction of earth station; θ_T: satellite antenna half beamwidth angle; G_{Rmax}: earth station antenna receive gain at boresight; θ_R: earth station antenna depointing angle; L_{FRX}: feeder loss from earth station antenna to receiver input; C_D: carrier power at receiver input; RX: receiver.

compared to its maximum. From Table 5.1, the ratio $(C/N_0)_D$ is defined as:

$$\left(\frac{C}{N_0}\right)_D = \frac{C_D}{N_{0D}} \quad \text{(Hz)} \tag{5.33}$$

where C_D is the power of the received carrier at the input to the earth station receiver. N_{0D} is the power spectral density of noise and relates to the downlink system noise temperature T_D:

$$N_{0D} = kT_D \quad \text{(W/Hz)} \tag{5.34}$$

where k is the Boltzmann constant: $k = 1.38 \times 10^{-23}$ J/K; $k(\text{dBJ/K}) = 10 \log k = -228.6$ dBJ/K.

$(C/N_0)_D$ can be expressed [MAR93, p. 65] as:

$$\left(\frac{C}{N_0}\right)_D = \text{OBO}_1 \left(\frac{C}{N_0}\right)_{\text{Dsat}} \quad \text{(Hz)}$$

$$\left(\frac{C}{N_0}\right)_D (\text{dBHz}) = \text{OBO}_1(\text{dB}) + \left(\frac{C}{N_0}\right)_{\text{Dsat}} (\text{dBHz}) \tag{5.35}$$

where OBO_1 is the input back-off per carrier for the considered carrier, as defined in Appendix 6 by expression (A6.4). OBO_1 depends on IBO_1. Simplified models are given by (A6.10) and (A6.11). $(C/N_0)_{\text{Dsat}}$ is the value of $(C/N_0)_D$ obtained at transponder saturation, and is given by:

$$\left(\frac{C}{N_0}\right)_{\text{Dsat}} = \text{EIRP}_{\text{SLsat}} \times \frac{1}{L_0} \times \left(\frac{G}{T}\right)_{\text{ES}} \times \frac{1}{k} \quad \text{(Hz)}$$

$$\left(\frac{C}{N_0}\right)_{\text{Dsat}} (\text{dBHz}) = \text{EIRP}_{\text{SLsat}}(\text{dBW}) - L_D(\text{dB}) + \left(\frac{G}{T}\right)_{\text{ES}} (\text{dBK}^{-1}) - 10 \log k(\text{dBJ/K})$$
$$\tag{5.36}$$

$\text{EIRP}_{\text{SLsat}}$ is the effective isotropic radiated power of the satellite (W) when operated at saturation (see section 5.3.1). L_D is the downlink path loss (see section 5.3.3). $(G/T)_{\text{ES}}$ is the figure of merit of the earth station (see section 5.3.4). k is the Boltzmann constant: $k = 1.38 \; 10^{-23}$ J/K; $k(\text{dBJ/K}) = 10 \log k = -228.6$ dBJ/K.

The above expressions assume that the satellite transmits a noise-free carrier so that the noise which corrupts the carrier at the earth station receiver is downlink noise only, with neither contribution from the uplink noise relayed by the transponder, nor from interference.

5.3.1   Effective isotropic radiated power of the satellite

From the definition given in Appendix 5, the effective isotropic radiated power of the satellite is expressed as:

$$\text{EIRP}_{\text{SL}} = P_T G_T \quad \text{(W)}$$

$$\text{EIRP}_{\text{SL}}(\text{dBW}) = P_T(\text{dBW}) + G_T(\text{dBi}) \tag{5.37}$$

where P_T is the power fed to the transmitting antenna, and G_T is the satellite antenna transmit gain along the constant gain coverage contour.

The power P_T fed to the antenna depends on the power P_{TX} at the output of the transponder amplifier and the feeder loss L_{FTX}:

$$P_T = \frac{P_{TX}}{L_{FTX}} \quad (W)$$

$$P_T(dBW) = P_{TX}(dBW) - L_{FTX}(dB) \tag{5.38}$$

$EIRP_{SLsat}$ is obtained when the transponder amplifier operates at saturation. The actual satellite EIRP depends on the total output back-off:

$$EIRP_{SL} = OBO_t EIRP_{SLsat} \quad (W)$$

$$EIRP_{SL}(dBW) = OBO_t(dB) + EIRP_{SLsat}(dBW) \tag{5.39}$$

OBO_t is a function of IBO_t, as discussed in Appendix 6.

Table 2.2 gives typical values of satellite transponder $EIRP_{SLsat}$ according to the type of coverage.

5.3.2 Flux density at earth surface

According to the definition given in Appendix 5, the flux density generated by satellite transmission at earth surface is equal to:

$$\Phi = \frac{EIRP_{SL}}{4\pi R^2} \quad (W/m^2)$$

$$\Phi(dBW/m^2) = EIRP_{SL}(dBW) - \log 4\pi R^2 \tag{5.40}$$

Radio regulations impose limits to the flux density at earth surface produced by satellites; according to expression (5.40), such limits translate into maximum values of $EIRP_{SL}$. The objective here is to limit the level of interference from an interfering satellite onto an earth station in the wanted satellite network. This aspect is discussed in more detail in section 5.5.

5.3.3 Downlink path loss

The downlink path loss L_D is the overall attenuation of the carrier power on its way from the satellite transmitting antenna to the earth station receiving antenna. As for the uplink, it builds up from two components, the free space loss, L_{FS}, and the atmospheric loss, L_A, so that the path loss can be expressed as:

$$L_D = L_{FS}L_A$$

$$L_D(dB) = L_{FS}(dB) + L_A(dB) \tag{5.41}$$

The downlink free space loss L_{FS} depends on the downlink frequency f and on the distance R between the earth station and the satellite, with expression identical to (5.28) used for the uplink free space loss. The distance R is taken into account by the same geometric factor $(R/R_0)^2$ as for the uplink, so expression (5.29) can be used as well.

Values can be obtained from Figures 5.13 and 5.14, using the appropriate downlink frequency, and relative coordinates of the receiving earth station.

The atmospheric loss L_A is again the result of the combined effect of attenuation due to atmospheric gases A_{AG} and attenuation due to rain A_{RAIN}. Values for A_{AG} have been discussed in the context of the uplink. Calculation of A_{RAIN} can be performed according to the above cited ITU References (Reports 563, 564, 721 and 723). The curves of Figures 5.16 and 5.17 can be considered for a preliminary evaluation of the value of A_{RAIN} exceeded for a given percentage of time depending on the climatic region in a European context.

5.3.4 Figure of merit of earth station receiving equipment

Figure 5.19 displays the components of the receiving equipment: the antenna, the feeder from antenna to receiver and the receiver. The receiver comprises a low noise amplifier (LNA), a down converter constituted of a mixer and a local

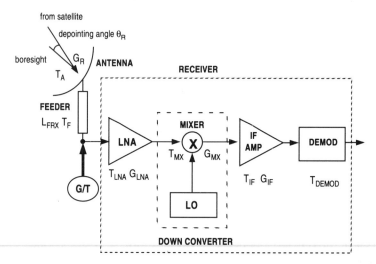

Figure 5.19 Earth station receiving equipment components. G_R: receive antenna gain in the direction of the satellite; T_A: antenna noise temperature; L_{FRX}: feeder loss; T_F: feeder temperature; LNA: low noise amplifier with effective input noise temperature T_{LNA} and power gain G_{LNA}; MIXER with effective input noise temperature T_{MX} and power gain G_{MX}; IF AMP: intermediate frequency amplifier with effective input noise temperature T_{IF} and power gain G_{IF}; DEMOD: demodulator with input noise temperature T_{DEMOD}.

oscillator (LO), an intermediate frequency amplifier (IF AMP) and a demodulator (DEMOD).

As can be seen from expressions (5.35) and (5.36), the figure of merit $(G/T)_{ES}$ of this equipment measures its ability to build up a high value of $(C/N_0)_D$. Its expression is given by:

$$\left(\frac{G}{T}\right)_{ES} = \left(\frac{G_{Rmax}}{L_R L_{pol}}\right)_{ES} \left(\frac{1}{L_{FRX}}\right)_{ES} \left(\frac{1}{T_D}\right) \quad (K^{-1})$$

$$\left(\frac{G}{T}\right)_{ES} (dBK^{-1}) = G_{Rmax}(dBi) - L_R(dB) - L_{pol}(dB) - L_{FRX}(dB) - 10\log T_D \quad (5.42)$$

G_{Rmax} is the earth station antenna receive gain at boresight. L_R is the off-axis gain loss corresponding to the depointing angle θ_R. L_{pol} is the gain loss corresponding to antenna polarisation mismatch, typically 0.1 dB. L_{FRX} is the feeder loss from the antenna to the receiver input, typically 0.5 dB. T_D is the downlink system noise temperature, given by:

$$T_D = \frac{T_A}{L_{FRX}} + T_F\left(1 - \frac{1}{L_{FRX}}\right) + T_R \quad (K) \tag{5.43}$$

T_A is the earth station antenna noise temperature. T_F is the temperature of the feeder, typically 290 K. T_R is the earth station receiver effective input noise temperature.

The earth station antenna receive gain at boresight is given by:

$$G_{Rmax} = \eta_a\left(\frac{\pi Df}{c}\right)^2 \quad (K^{-1})$$

$$G_{Rmax}(dBi) = 10\log[\eta_a(\pi Df/c)^2] \tag{5.44}$$

Values of G_{Rmax} are given in Figures 5.7 and 5.8, where the value of frequency to be considered is that of the downlink frequency.

The actual receive gain in the direction of the satellite is:

$$G_R(dBi) = G_{Rmax}(dBi) - L_R(dB) - L_{pol}(dB) \tag{5.45}$$

where the off-axis receive gain loss L_R corresponding to the depointing angle θ_R is given by:

$$L_R(dB) = 12\left(\frac{\theta_R}{\theta_{3dB}}\right)^2 \tag{5.46}$$

where θ_{3dB} is the half power beamwidth of the receive radiation pattern as given by expression (5.16) and curves of Figures 5.7 and 5.8, considering the actual value of the downlink frequency.

The determination of the maximum value θ_{Rmax} of θ_R has been discussed in section 5.2.2 and is illustrated in Figure 5.10. Actually, θ_{Rmax} is equal to θ_{Tmax} as given by expression (5.25). Values of $L_{Rmax}(dB)$ are readily calculated from (5.24) replacing θ_T by θ_R and considering $\theta_R = \theta_{Rmax}$.

Expression (5.43) shows that the downlink system noise temperature depends on the earth station antenna noise temperature T_A. The antenna noise temperature T_A represents the overall contribution of noise components captured by the antenna. Two situations are to be considered:

—Antenna noise temperature for *clear sky conditions* (Figure 5.20). The antenna captures the noise radiated by the sky with temperature T_{SKY} and a contribution T_{GROUND} from the ground in the vicinity of the earth station. The overall contribution is given by:

$$T_A = T_{SKY} + T_{GROUND} \quad (K) \tag{5.47}$$

—Antenna noise temperature for *rain conditions* (Figure 5.21): Rain acts as an attenuator with attenuation A_{RAIN} and average medium temperature T_m (typically $T_m = 278\,K$). It attenuates the contribution from the clear sky, and generates its own noise with noise temperature $T_m(1 - 1/A_{RAIN})$ at the output of the attenuation process. The noise contribution from the ground in the vicinity of the earth station is considered not to be modified by rain. The overall contribution is given by:

$$T_A = \frac{T_{SKY}}{A_{RAIN}} + T_m\left(1 - \frac{1}{A_{RAIN}}\right) + T_{GROUND} \quad (K) \tag{5.48}$$

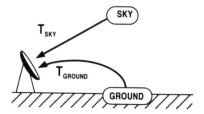

Figure 5.20 Components of earth station antenna noise temperature for clear sky conditions.

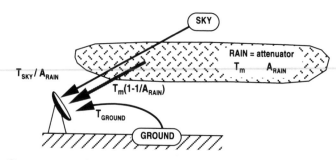

Figure 5.21 Components of earth station antenna noise temperature for rain conditions.

Table 5.3 Clear sky noise contribution T_{SKY} (standard atmosphere)

Frequency	$E = 10°$	$E = 35°$
4 GHz	10 K	4 K
12 GHz	20 K	7 K

Table 5.3 displays typical values of the clear sky noise T_{SKY} at two elevation angles ($E = 10°$ and $E = 35°$) for a standard atmosphere. The contribution of the ground in the vicinity of the earth station depends on the type of antenna (mounting, diameter), the side lobe radiation pattern and the frequency. A typical value is $T_{GROUND} = 30$ K.

Figure 5.22 displays the variation of the antenna noise temperature T_A with rain attenuation A_{RAIN}. The antenna noise temperature increases with rain attenuation.

The earth station receiver effective input noise temperature T_R can be calculated from Friis' formula:

$$T_R = T_{LNA} + \frac{T_{MX}}{G_{LNA}} + \frac{T_{IF}}{G_{LNA}G_{MX}} + \frac{T_{DEMOD}}{G_{LNA}G_{MX}G_{IF}} \quad (K) \qquad (5.49)$$

In the above formula, all values of noise temperature and gain are to be expressed in absolute values (not in dB). Usually, the LNA gain is high enough (typically 50 dB $= 10^5$) for all terms but the first one to be negligible. Therefore $T_R \approx T_{LNA}$, with typical value equal to 30 K at C-band and 80 K at Ku-band.

From the above set of expressions, one can see that the figure of merit $(G/T)_{ES}$ of an earth station is maximum when there is no depointing loss, no feeder

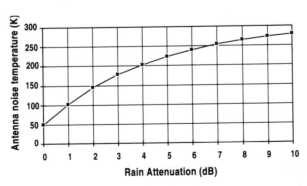

Figure 5.22 Variation of antenna noise temperature with rain attenuation (frequency 12 GHz, $T_{SKY} = 20$ K, $T_{GROUND} = 30$ K, $T_m = 278$ K).

loss, no polarisation mismatch and no rain attenuation. This maximum value is given by:

$$\left(\frac{G}{T}\right)_{ESmax} = \frac{(G_{Rmax})_{ES}}{T_{Dmin}} \quad (K^{-1})$$

$$\left(\frac{G}{T}\right)_{ESmax} (dBK^{-1}) = G_{Rmax}(dBi) - 10 \log T_{Dmin} \quad (5.50)$$

where:

$$T_{Dmin} = T_{SKY} + T_{GROUND} + T_R \quad (K) \quad (5.51)$$

Figures 5.23 and 5.24 display variations of $(G/T)_{ESmax}$ with antenna diameter respectively at 4 GHz (C-band) and 12 GHz (Ku-band), considering the receiver noise temperature T_R as a parameter.

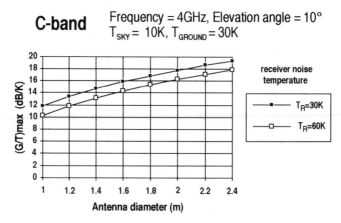

Figure 5.23 Variations of $(G/T)_{ESmax}$ with antenna diameter at 4 GHz.

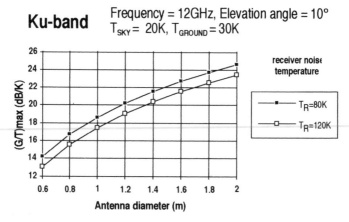

Figure 5.24 Variations of $(G/T)_{ESmax}$ with antenna diameter at 12 GHz.

When a more representative value is sought, the impacts of depointing and feeder losses are to be introduced along with the impact of rain attenuation. The resulting degradation of $(G/T)_{ES}$ with respect to $(G/T)_{ESmax}$ can be evaluated from:

$$\left(\frac{G}{T}\right)_{ES} (dBK^{-1}) = \left(\frac{G}{T}\right)_{ESmax} (dBK^{-1})$$

$$- L_{Rmax}(dB) \qquad \text{depointing loss}$$

$$- L_{pol}(dB) \qquad \text{polarisation mismatch loss}$$

$$- \text{DELTA}(dB) \qquad \text{combined effect of feeder loss and} \qquad \text{rain attenuation} \qquad (5.52)$$

where:

$$L_{Rmax}(dB) = 12\left(\frac{\theta_{Rmax}}{\theta_{3dB}}\right)^2 \quad (dB) \qquad (5.53)$$

$$L_{pol} = \text{typically } 0.1 \text{ dB}$$

$$\text{DELTA}(dB) = L_{FRX} + 10\log T_D - 10\log T_{Dmin} \qquad (5.54)$$

In expression (5.54), T_D represents the downlink system noise temperature as given by (5.43), with T_A given by (5.48).

For a VSAT, typical values of L_{Rmax} are 0.6 dB at C-band and 1 dB at Ku-band. Tables 5.4 and 5.5 display typical values of DELTA.

From the above results, it is possible to work out Tables 5.6 and 5.7 which indicate achievable values of G/T for VSATs depending on antenna diameter, considering an elevation angle $E = 35°$. Table 5.8 indicates typical values for hub stations.

Table 5.4 Typical values of DELTA at 4 GHz (C-band): elevation angle$=35°$; $T_{SKY}=4$ K; $T_{GROUND}=30$ K; feeder loss $L_{FRX}=0.5$ dB; station keeping window half-width SKW $= 0.025°$

C-band		A_{RAIN}	
		0 dB	0.6 dB*
T_R	30 K	2.1 dB	3.3 dB
	60 K	1.6 dB	2.6 dB

*0.6 dB corresponds to the value of attenuation due to rain which is exceeded for 0.01% of the time in region H of Europe (see Figure 5.16).

Table 5.5 Typical values of DELTA at 12 GHz (Ku-band): elevation angle $= 35°$; $T_{SKY} = 7$ K; $T_{GROUND} = 30$ K; feeder loss $L_{FRX} = 0.5$ dB; station keeping window half-width SKW $= 0.025°$

Ku-band		A_{RAIN}	
		0 dB	6 dB*
T_R	80 K	1.4 dB	4.9 dB*
	120 K	1.2 dB	4.1 dB

* 6 dB corresponds to the value of attenuation due to rain which is exceeded for 0.01% of the time in region H of Europe (see Figure 5.17).

Table 5.6 Achievable *G/T* with current VSAT technology at C-band (elevation angle $E = 35°$)

	C-band	
Antenna diameter	1.8 m	2.4 m
G/T clear sky	15 dBK^{-1}	17 dBK^{-1}
G/T rain (0.01%)*	13 dBK^{-1}	16 dBK^{-1}

*Europe region H.

Table 5.7 Achievable *G/T* with current VSAT technology at Ku-band (elevation angle $E = 35°$)

	Ku-band	
Antenna diameter	1.2 m	1.8 m
G/T clear sky	19 dBK^{-1}	22 dBK^{-1}
G/T rain (0.01%)*	15 dBK^{-1}	19 dBK^{-1}

*Europe region H.

Table 5.8   Typical G/T for hub station (elevation angle $E = 35°$)

	C-band		
Antenna diameter	$3\,m^\dagger$	$6\,m^\dagger$	$10\,m$
G/T clear sky	$19\,dBK^{-1}$	$25\,dBK^{-1}$	$29\,dBK^{-1}$
G/T rain (0.01%)*	$18\,dBK^{-1}$	$23\,dBK^{-1}$	$28\,dBK^{-1}$

	Ku-band		
Antenna diameter	$3\,m^\dagger$	$6\,m$	$10\,m$
G/T clear sky	$26\,dBK^{-1}$	$33\,dBK^{-1}$	$37\,dBK^{-1}$
G/T rain (0.01%)*	$22\,dBK^{-1}$	$29\,dBK^{-1}$	$34\,dBK^{-1}$

*Europe region H.
†No tracking.

5.4   INTERMODULATION ANALYSIS

When operated in a non-linear mode, the transponder amplifier generates inter-modulation products, and the transponder output power is shared not only between the amplified carriers but also with the intermodulation products. This is most pronounced at and near transponder saturation, when the total input back-off is equal to, or approaches, zero (see Appendix 6 for definition of back-off). Intermodulation products of modulated carriers are transmitted along with the wanted carrier and act as noise for the wanted carrier as received by the destination earth station.

Figure 5.25 gives typical curves for the carrier power to intermodulation noise power density ratio $(C/N_0)_{IM}$ at the destination earth station receiver input, as a function of the transponder amplifier total input back-off, assuming n equally powered carriers.

An approximate formula for the curves of Figure 5.25 is [ITU88, p. 90]:

$$\left(\frac{C}{N_0}\right)_{IM} = 79 - 10 \log n - 1.65(IBO_t(dB) + 5) \quad IBO_t < -5\,dB \quad (dBHz) \quad (5.55)$$

For the EUTELSAT 2 satellite, the following formula can be used [EUT92]:

$$\left(\frac{C}{N_0}\right)_{IM} = 84.2 - 0.34\,IBO_t(dB) + 0.02(IBO_t(dB))^2 + 10 \log B/36 - 10 \log n \quad (dBHz)$$

$$\text{for } -13\,dB < IBO_t < 0\,dB \quad (5.56)$$

where B is the transponder bandwidth occupied by the n equally powered carriers.

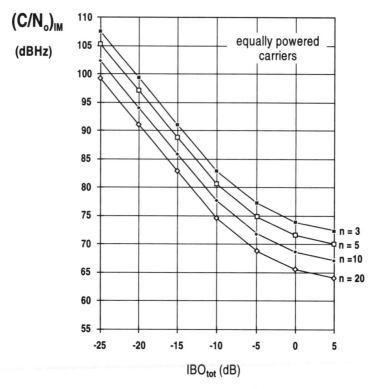

Figure 5.25 $(C/N_0)_{IM}$ as a function of the total input back-off.

5.5 INTERFERENCE ANALYSIS

5.5.1 Expressions for carrier-to-interference ratio

Interference is unwanted radio frequency energy introduced at the end receiver of the wanted link. Figure 5.26 illustrates the spectrum of a wanted carrier contaminated by an interfering carrier. The receiver with equivalent noise bandwidth B_N captures carrier power C, and interference power N_i. The carrier-to-interference power ratio is C/N_i.

The amount of interference power into the receiver depends on the overlap of the carrier sprectra, as illustrated in Figure 5.26.

The receiver noise bandwidth B_N is matched to the bandwidth occupied by the wanted carrier with power C. Power C is given by:

$$C = PSD_w B_N \quad (W) \qquad\qquad (5.57)$$

where PSD_w is the wanted carrier power spectral density (W/Hz).

The interfering carrier has a power spectral density PSD_i (W/Hz) and occupies a band B_i. The amount of interfering power N_i captured by the receiver is

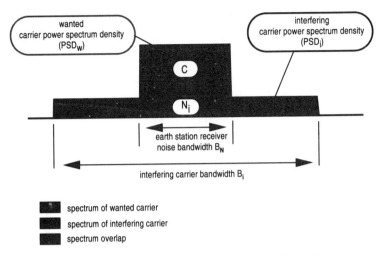

Figure 5.26 Spectrum of interfering carrier bandwidth B_i

spectrum of wanted carrier

spectrum of interfering carrier

spectrum overlap

Figure 5.26 Spectrum of received carrier contaminated by an interfering carrier.

equal to the spectrum overlap area:

$$N_i = PSD_i \min(B_i, B_N) \quad (W) \tag{5.58}$$

where $\min(B_i, B_N)$ represents the smallest of the two bandwidths, B_i or B_N.

The expression for the carrier-to-interference power ratio C/N_i results from (5.57) and (5.58):

$$\frac{C}{N_i} = PSD_w B_N / PSD_i \min(B_i, B_N) \tag{5.59}$$

Carrier power spectral density, PSD, is proportional to the EIRP density defined as EIRP/B where B is the carrier bandwidth (Hz). Indeed:

$$PSD = \frac{EIRP}{B} \frac{1}{L} G_{RX} \quad (W/Hz) \tag{5.60}$$

where:

L is the satellite link path loss,
G_{RX} is the composite receive gain expressed as $G_{RX} = G_{Rmax}/L_R L_{pol} L_{FRX}$ (5.61)
G_{Rmax} is the antenna receive gain at boresight,
L_R is the off-axis gain loss,
L_{pol} is the gain loss corresponding to antenna polarisation mismatch,
L_{FRX} is the feeder loss from antenna to receiver input.

The above derivations are approximate as no account is made of the actual shape of the modulated carrier spectrum. A more realistic approach should also consider that the centre frequency of the spectrum of the wanted carrier and the interfering carrier may not be coincident.

5.5.2   Types of interference

Interference can be:

—*self-interference*, i.e. transmissions by the wanted satellite system which hosts the considered VSAT network; or

—*external interference*, i.e. transmissions from systems sharing the same frequency bands as the wanted satellite system. Candidate interfering systems are other satellite systems and terrestrial microwave systems. C-band and most of Ku-band are shared by satellite and terrestrial microwave systems (see Figure 1.15). It is therefore convenient to operate VSAT networks in bands that are allocated to satellite links on a primary and exclusive basis, as this eliminates the interference from terrestrial microwave systems. However, this does not protect a VSAT network from being interfered with by other satellite-based networks.

5.5.3   Self-interference

Self-interference is caused by frequency reuse and imperfect filtering. The first type is often called *co-channel interference* (CCI), the second type *adjacent channel interference* (ACI).

5.5.3.1   *Co-channel interference*

Co-channel interference (CCI) is a drawback of frequency reuse techniques which aim at increasing the capacity of a satellite system without demanding more frequency band. These techniques have been presented in Chapter 2, section 2.1.5.

Co-channel interference originates in imperfect isolation between geographically separated beams or between orthogonal polarisations reusing the same frequency band. Therefore, it strongly depends on the antenna characteristics of the earth station and the satellite.

Beam to beam interference.   Figure 5.27 illustrates the generation of co-channel interference as a result of imperfect isolation between beams in a multibeam satellite system. The VSAT network is assumed to be located within beam 1. An earth station belonging to some other network is located in beam 2. As a result of the non-zero gain of the antenna gain pattern defining beam 1 in the direction of this earth station, some of its emitted power (upward arrow in black) is received in beam 1 (upward arrow in grey). This power enters the transponder conveying the inbound and outbound links of the VSAT network. As frequency reuse implies that both beams use the same frequency band, there is a chance of carrier spectrum overlap between the carrier emitted by the station of beam 2 and one or several of the VSAT network uplinked carriers (bidirectional arrows in black representing inbound and outbound links).This overlap constitutes uplink interference into the VSAT network.

The transponder dedicated to beam 2 transmits carriers to stations of beam 2. The downward arrow in black in beam 2 represents one of these carriers. A part

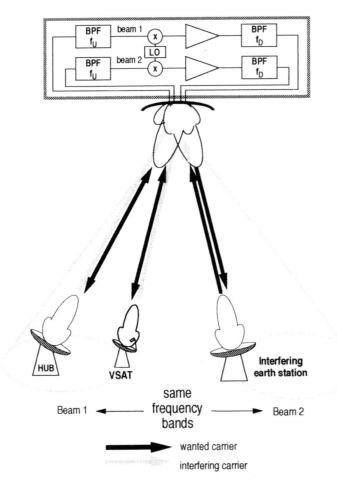

Beam 1 ←————— same frequency bands —————→ Beam 2

————————▶ wanted carrier

⋯⋯⋯⋯▶ interfering carrier

Figure 5.27 Generation of co-channel interference in a multibeam satellite system. BPF = band pass filter, LO = local oscillator

of its power is emitted in beam 1 (downward arrows in grey in beam 1) and the part of its spectrum that overlaps with the spectrum of either the oubound or any of the inbound carriers of the VSAT network constitutes downlink interference to the VSAT network.

Expressions for carrier-to-co-channel interference power ratio C/N_i can be derived from (5.59) and (5.60):

Uplink co-channel interference:

$$\left(\frac{C}{N_i}\right)_U = \frac{\text{EIRP}_{\text{ESw}}}{\text{EIRP}_{\text{ESi}}} \frac{B_i}{\min[B_i, B_N]} \frac{L_{Ui}}{L_{Uw}} \frac{G_{\text{RXw}}}{G_{\text{RXi}}}$$

$$\left(\frac{C}{N_i}\right)_U (\text{dB}) = \text{EIRP}_{\text{ESw}}(\text{dBW}) - \text{EIRP}_{\text{ESi}}(\text{dBW}) + 10\log B_i - 10\log \min[B_i, B_N]$$

$$+ L_{Ui}(\text{dB}) - L_{Uw}(\text{dB}) + G_{\text{RXw}}(\text{dBi}) - G_{\text{RXi}}(\text{dBi}) \tag{5.62}$$

where EIRP_{ESw} is the EIRP of the wanted carrier of beam 1 in the direction of the satellite; EIRP_{ESi} is the EIRP of the interfering carrier of beam 2 in the direction of the satellite; L_{Uw} and L_{Ui} are the uplink path loss for the wanted carrier and the interfering one, respectively; and G_{RXw} and G_{RXi} are the satellite composite receive gain of beam 1, given by (5.61), for the wanted and interfering carriers, respectively. The worst case should consider that the receiving station is at edge of beam 1 and the interfering station at edge of beam 2, as near as possible to the receiving station, so that the difference $G_{RXw}(\text{dBi}) - G_{RXi}(\text{dBi})$ is minimum. The difference value can be obtained from the satellite coverage charts, or if not available, from the reference satellite antenna gain patterns given in the ITU-R Recommendation 672 and Report 558.

Downlink co-channel interference:

$$\left(\frac{C}{N_i}\right)_D = \frac{\text{EIRP}_{SL1ww}}{\text{EIRP}_{SL2iw}} \frac{B_i}{\min[B_i, B_N]}$$

$$\left(\frac{C}{N_i}\right)_D (\text{dB}) = \text{EIRP}_{SL1ww}(\text{dBW}) - \text{EIRP}_{SL2iw}(\text{dBW}) + 10\log B_i - 10\log\min[B_i, B_N]$$

$$(5.63)$$

where EIRP_{SL1ww} is the satellite EIRP in beam 1 for the wanted carrier in the direction of the receiving (wanted) station. Its value depends on the output back-off for that carrier, OBO_w, and the satellite EIRP in beam 1 at saturation in the direction of the wanted station, $\text{EIRP}_{SL1wsat}$:

$$\text{EIRP}_{SL1ww}(\text{dBW}) = \text{OBO}_w(\text{dB}) + \text{EIRP}_{SL1wsat}(\text{dBW}) \qquad (5.64)$$

EIRP_{SL2iw} is the satellite EIRP in beam 2 for the interfering carrier in the direction of the receiving (wanted) station. Its value depends on the output back-off for that carrier, OBO_i, and the satellite EIRP in beam 2 at saturation, $\text{EIRP}_{SL2wsat}$ in the direction of the wanted station:

$$\text{EIRP}_{SL2iw}(\text{dBW}) = \text{OBO}_i(\text{dB}) + \text{EIRP}_{SL2wsat}(\text{dBW}) \qquad (5.65)$$

The worst case should consider that the wanted station is at edge of beam 1 and the interfering station at edge of beam 2, as near as possible to the wanted station so that the difference $\text{EIRP}_{SL1ww}(\text{dBW}) - \text{EIRP}_{SL2iw}(\text{dBW})$ is minimum.

If one assumes that beams 1 and 2 have equal saturated maximum EIRP and identical gain pattern, then:

$$\text{EIRP}_{SL1wsat}(\text{dBW}) = \text{EIRP}_{SL2wsat}(\text{dBW}) + G_{T1w}(\text{dBi}) - G_{T2w}(\text{dBi}) \qquad (5.66)$$

where $G_{T1w}(\text{dBi})$ and $G_{T2w}(\text{dBi})$ represents respectively the transmit antenna gain in the direction of the wanted station for the common antenna gain pattern of beam 1 and 2.

Combining expressions (5.63) to (5.66):

$$\left(\frac{C}{N_i}\right)_D (\text{dB}) = \text{OBO}_w(\text{dB}) - \text{OBO}_i(\text{dB}) + 10\log B_i - 10\log\min[B_i, B_N]$$

$$+ G_{T1w}(\text{dBi}) - G_{T2w}(\text{dBi}) \qquad (5.67)$$

Example

Consider a multibeam satellite with the same EIRP and antenna pattern in all beams. Table 5.9 displays the parameters for a wanted VSAT, the interfering station, and satellite relevant characteristics, and gives results of $(C/N_i)_U$ and $(C/N_i)_D$ calculation.

Overall link co-channel interference:

Co-channel interference may contaminate both the uplink and the downlink portions of the VSAT network inbound or outbound link. The overall link carrier-to-co-channel interference power ratio is given by:

$$\left(\frac{C}{N_i}\right)_T^{-1} = \left(\frac{C}{N_i}\right)_U^{-1} + \left(\frac{C}{N_i}\right)_D^{-1} \qquad (5.68)$$

where the absolute values of $(C/N_i)_U$ and $(C/N_i)_D$, calculated according to expressions (5.62) and (5.67) respectively, should be used.

Table 5.9   Carrier-to-co-channel interference power ratio calculation

UPLINK					
	VSAT	EIRP	1	$EIRP_{ESw}$	40 dBW
		Receiver noise equivalent bandwidth	2	B_N	100 kHz = 50 dBHz
		Uplink path loss	3	L_{Uw}	207 dB
	Interfering station	EIRP	4	$EIRP_{ESi}$	65 dBW
		Interfering carrier bandwidth	5	B_i	2 MHz = 63 dBHz
		Uplink path loss	6	L_{Ui}	207 dB
	Satellite	Gain pattern	7	$G_{RXw} - G_{RXi}$	30 dB
	$(C/N_i)_U$		8	Expression (5.62) $= 1 - 2 - 3 - 4 + 5 + 6 + 7$	18 dB
DOWNLINK					
	Satellite	Wanted carrier output back-off	9	OBO_w	-30 dB
		Interfering carrier output back-off	10	OBO_i	-5 dB
		Antenna gain difference	11	$G_{T1w} - G_{T2w}$	30 dB
	$(C/N_i)_D$		12	Expression (5.67) $= 9 - 10 + 5 - 2 + 11$	18 dB

For any link, the carrier power-to-co-channel interference power spectral density ratio (C/N_{0i}) is given by:

$$\frac{C}{N_{0i}} = \frac{C}{N_i} \times B_N \quad (Hz)$$

$$\frac{C}{N_{0i}}(dBHz) = \frac{C}{N_i}(dB) + 10 \log B_N \qquad (5.69)$$

Cross polarisation interference. Co-channel interference is also caused by orthogonally polarised carriers reusing the same frequency band within a given beam. Should the interfering stations be part of the considered VSAT network, then the VSAT network operation requires the use of two transponders, one for each group of carriers with the same polarisation. This is generally not justified for VSAT networks, hence all VSATs and the hub within the geographical coverage of the beam operate on the same polarisation. Cross polarisation interference then originates in carriers transmitted by stations of other networks using the same satellite system and operating within the same frequency bands as the considered

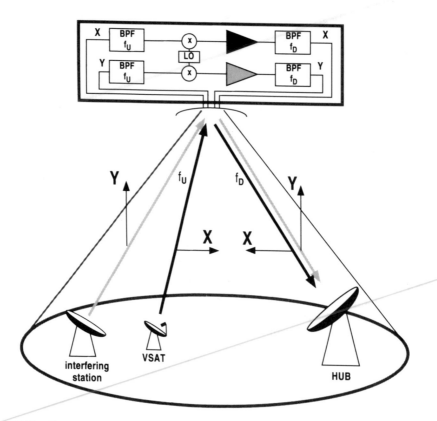

Figure 5.28 Frequency reuse based on transmission of two carriers at the same frequency with orthogonal polarisations X and Y; BPF: band pass filter.

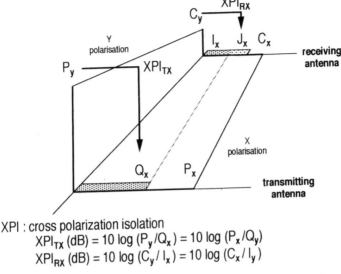

XPI : cross polarization isolation
$$XPI_{TX} (dB) = 10 \log (P_y/Q_x) = 10 \log (P_x/Q_y)$$
$$XPI_{RX} (dB) = 10 \log (C_y/I_x) = 10 \log (C_x/I_y)$$

Figure 5.29 Contribution of energy on the horizontal component from the vertical one at the transmitting and receiving ends of a satellite link. TX: transmitter, RX: receiver

VSAT network but on the orthogonal polarisation. These stations may be located within the same beam, and their carriers are then conveyed by the cross-polar satellite transponder, as illustrated in Figure 5.28.

The level of interference is conditioned by the cross polarisation isolation (XPI) of earth station and satellite antennas. The cross polarisation of an antenna is defined in Appendix 4. Figure 5.29 illustrates how energy from one polarisation contaminates the orthogonal one altogether at the transmitting and the receiving ends of a satellite link.

Carriers are transmitted at the same frequency, one on each X and Y polarisation. P_x and P_y are the respective transmitted powers. Due to imperfect cross polarisation isolation, a contribution Q_x is transmitted along with P_x. The receiving antenna receives the wanted carrier C_x to which adds I_x, the interfering contribution from Q_x. On the orthogonal polarisation, it receives C_y, and due to imperfect cross polarisation isolation, a contribution J_x is generated, which adds to I_x, so the total interference power is $I_x + J_x$. A similar process occurs on the Y-component. Table 5.10 displays the different power components and the resulting carrier to interference ratio.

From Table 5.10, an expression for the carrier-to-interference power ratio can be derived:

$$\frac{C_x}{I_x + J_x} = \left[\frac{P_y}{P_x} 10^{-\frac{XPI_{TX}}{10}} + \frac{C_y}{C_x} 10^{-\frac{XPI_{RX}}{10}} \right]^{-1} \tag{5.70}$$

$$\frac{C_y}{I_y + J_y} = \left[\frac{P_x}{P_y} 10^{-\frac{XPI_{TX}}{10}} + \frac{C_x}{C_y} 10^{-\frac{XPI_{RX}}{10}} \right]^{-1} \tag{5.71}$$

where values for XPI are in dB.

Table 5.10 Components of wanted and interference carrier power components and resulting carrier-to-cross-polarisation interference power ratio (enter values for XPI in dB)

	X-polarisation	Y-polarisation
Transmitted power (W)	P_x	P_y
Cross polar transmitted power (W)	$Q_x = P_y \times 10^{-(\text{XPI}_{\text{TX}}/10)}$	$Q_y = P_x \times 10^{-(\text{XPI}_{\text{TX}}/10)}$
Received carrier power (W)	C_x	C_y
Generated cross polar interference (W)	$I_x = C_y \times 10^{-(\text{XPI}_{\text{RX}}/10)}$	$I_y = C_x \times 10^{-(\text{XPI}_{\text{RX}}/10)}$
Received co-polar interference (W)	$J_x = Q_x C_x / P_x$	$J_y = Q_y C_y / P_y$
Carrier-to-cross-polarisation interference power ratio	$C_x/(I_x + J_x)$	$C_y/(I_y + J_y)$

Example:

Assuming the same power transmitted on both polarisations, $P_x = P_y$, the same path loss for both polarisations, $C_x = C_y$, and the same cross polarisation isolation for the transmitting and the receiving antenna, $\text{XPI}_{\text{TX}} = \text{XPI}_{\text{RX}} = 27\,\text{dB}$, then the carrier-to-cross-polarisation interference power ratio for the considered link is:

$$\frac{C}{N_i} = \frac{C_x}{I_x + J_x} = \frac{C_y}{I_y + J_y} = \{10^{-2.7} + 10^{-2.7}\}^{-1} = 250.6 = 24\,\text{dB}$$

However, it may be that the power transmitted by the interfering station is much higher than the power of a transmitting VSAT. Assume, for instance, that the interfering station transmits on the Y-polarisation with an EIRP 10 dB higher than that of a VSAT on its inbound X-polarised link, then $P_y/P_x = C_y/C_x = 10$ and C/N_i reduces to 14 dB.

Similar considerations apply to the downlink.

The above calculations do not include depolarisation of the wave on its way from the transmitted to the receiver, which occurs in the atmosphere in the presence of rain, or ice clouds. Rain induced depolarisation results from differential attenuation and differential phase shift between two characteristic orthogonal polarisations. These effects originate in the non-spherical shape of rain drops.

The relationship between cross-polarised discrimination XPD (see Appendix 4 for definition) and the copolarised path attenuation A_{RAIN} is important for predictions based on attenuation statistics (as given by Figures 5.16 and 5.17). The following relationship is in agreement with long term measurements in the frequency range between about 3 GHz and 37 GHz (ITU-R Report 722):

$$\text{XPD(dB)} = 30\log f(\text{GHz}) - V\log A_{\text{RAIN}}(\text{dB}) - 40\log\cos E(\text{degrees}) + 0.0053\sigma^2 + I(\tau)$$

$$(5.72)$$

where:

$V = 20$ for $3\,\text{GHz} < f < 15\,\text{GHz}$,
$V = 23$ for $15\,\text{GHz} < f < 37\,\text{GHz}$,
E is the elevation angle in degrees,
σ is the effective standard deviation for raindrop canting angle distribution given by $\sigma = -5\log p$, where p is the considered time percentage when the value considered for A_{RAIN} is exceeded,
τ is the polarisation tilt angle relative to the horizontal for linear polarisation. $I(\tau)$ can be omitted for circular polarisation. It represents the improvement of linear polarisation with respect to circular polarisation. If the effective canting angle is assumed to vary randomly within a rainstorm and from storm to storm and to have a Gaussian distribution with zero mean and standard deviation σ_m, then $I(\tau)$ can be expressed by:

$$I(\tau) = -10\log\{0.5[1 - \cos(4\tau)\exp(-0.0024\sigma_m)]\} \tag{5.73}$$

where σ_m in given in degrees.

Snow (dry or wet) causes similar phenomena. Ice clouds, where high altitude ice crystals are in a region close to the 0°C isotherm, also contribute to cross-polarisation. However, in contrast to rain and other hydrometeors, this effect is not accompanied by attenuation. It causes a reduction in the value of XPD given by (5.72) by an amount [ROG86]:

$$C_{\text{ice}} = (0.3 + 0.1\log p)\text{XPD} \quad (\text{dB}) \tag{5.74}$$

When all sources of depolarisation are included and assuming equally powered carriers on both polarisations, the carrier-to-cross-polarisation interference power ratio for a given link (uplink or downlink) can be calculated as:

$$\frac{C}{N_i} = \Delta\left\{10^{-\frac{\text{XPI}_{\text{TX}}}{10}} + 10^{-\frac{\text{XPD}}{10}} + 10^{-\frac{\text{XPI}_{\text{RX}}}{10}}\right\}^{-1} \tag{5.75}$$

where $\Delta = C_x/C_y$ is the ratio of the co-polar wanted carrier power to the cross-polar interfering carrier power (assuming the same path loss for both carriers), and XPI and XPD values are in dB.

The overall link (from station to station) carrier-to-cross-polarisation interference ratio $(C/N_i)_\text{T}$ is obtained by adding uplink and downlink interference power at the earth station receiver input:

$$\left(\frac{C}{N_i}\right)_\text{T}^{-1} = \left(\frac{C}{N_i}\right)_\text{U}^{-1} + \left(\frac{C}{N_i}\right)_\text{D}^{-1} \tag{5.76}$$

where the absolute values of $(C/N_i)_\text{U}$ and $(C/N_i)_\text{D}$, each calculated according to expression (5.75), should be used.

For any link, the carrier power-to-cross-polarisation interference power spectral density ratio (C/N_{0i}) is given by:

$$\frac{C}{N_{0i}} = \frac{C}{N_i} B_N \quad (Hz)$$

$$\frac{C}{N_{0i}} (dBHz) = \frac{C}{N_i} (dB) + 10\log B_N \tag{5.77}$$

5.5.3.2 Adjacent channel interference

Adjacent channel interference (ACI) originates in a part of the power of a carrier adjacent to a given carrier being captured by a satellite transponder or an earth

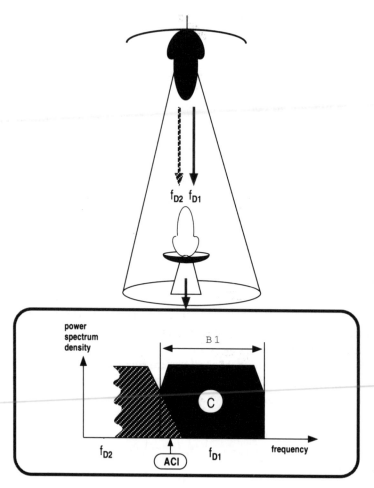

Figure 5.30 Downlink adjacent channel interference (ACI).

station receiver tuned to the frequency of the carrier considered. Figure 5.30 illustrates the situation where ACI affects a downlink carrier. Part of the spectrum of the adjacent carrier at frequency f_{D2} falls into the earth station receiver tuned to the carrier at frequency f_{D1}. The corresponding carrier-to-interference power ratio is determined by the area labelled C with respect to the area labelled ACI. Typical values of C/ACI for VSAT networks are in the range 25 to 30 dB. Adjacent channel interference can be reduced by adopting larger guard bands between carriers, at the expense of a larger utilised bandwidth on the satellite transponder.

5.5.4 External interference

External interference originates outside the considered satellite system: transmissions by other satellites and stations, be it earth stations or terrestrial ones (microwave relays on the Earth). Figure 1.17 illustrates the geometry of external interference. Actually, VSAT networks operating at Ku-band can be operated in the exclusive bands of Figure 1.15, and this protects those networks from microwave relay interference.

5.5.4.1 *Interference from adjacent satellite systems*

The interference generated by an earth station into an adjacent satellite or by an adjacent satellite into an earth station depends on their respective antenna radiation patterns. Actual antenna patterns should be considered but they may not be known at the early planning stages of a VSAT network. Therefore, reference patterns may be used instead. For VSATs and hub station antennas, such reference patterns are available from the ITU-R texts and VSAT standards presented in Chapter 1, section 1.9. For satellite antennas, one can refer to the ITU-R Recommendation 672 and Report 558. All reference patterns specify a sidelobe envelope level of the form $A - B \log \theta$ (dBi) where θ is the off-axis angle. One can refer, for example, to the reference pattern of a VSAT antenna in Table 1.11, where $A = 29$ dBi for $2.5° \leqslant \theta < 7°$ and $A = 32$ dBi for $9.2° \leqslant \theta < 48°$, and $B = 25$. Figure 5.31 displays a typical VSAT antenna radiation pattern and the reference sidelobe pattern of Table 1.11.

In order to calculate the level of interference encountered at the receiver end of a satellite link, it is necessary to know the angular separation between two geostationary satellites as seen by an earth station. Figure 5.32 displays the geometry where the satellites are separated in longitude by an angle α. The angle θ represents the angular separation between the two geostationary satellites as seen by the earth station.

The distance d between the two satellites is given by:

$$d^2 = R_w^2 + R_a^2 - 2R_w R_a \cos \theta \quad (m^2)$$

(5.78)

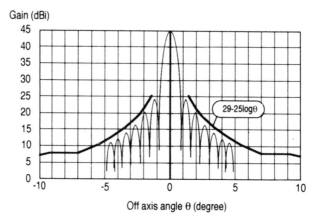

Gain (dBi)

Off axis angle θ (degree)

Figure 5.31 Typical VSAT antenna radiation pattern and sidelobe reference pattern of Table 1.11.

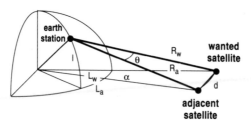

Figure 5.32 Geometry of earth station with respect to the wanted and the adjacent satellite.

where:

R_w is the slant range from the earth station to the wanted satellite,
R_a is the slant range from the earth station to the adjacent satellite.

On the other hand, $d/2 = (R_e + R_0) \sin(\alpha/2)$, resulting in:

$$d^2 = 4(R_e + R_0)^2 \sin^2 \frac{\alpha}{2} = 2(R_e + R_0)^2 (1 - \cos \alpha) \quad (m^2) \qquad (5.79)$$

where:

R_e is the earth radius: $R_e = 6378$ km,
R_0 is the satellite altitude: $R_0 = 35786$ km.

Combining (5.78) and (5.79) leads to:

$$\theta = \arccos \left[\frac{R_w^2 + R_a^2 - 2(R_e + R_0)^2 (1 - \cos \alpha)}{2 R_w R_a} \right] \qquad (5.80)$$

R_w and R_a are given by expression (2.8), and depend on the latitude l of the earth station and the relative longitude L_w and L_a with respect to the wanted and the adjacent satellite respectively. Note that $L_a - L_w = \alpha$. Table 5.11 provides some

Table 5.11 θ/α ratio for $\alpha = 4°$ orbital separation and various values of the earth station latitude l and relative longitude L_w with respect to the wanted satellite

	Relative longitude L_w		
Latitude l	$0°$	$10°$	$20°$
$0°$	1.18	1.17	1.16
$10°$	1.17	1.17	1.15
$20°$	1.16	1.16	1.15
$30°$	1.15	1.14	1.13
$40°$	1.12	1.12	1.11
$50°$	1.10	1.10	1.09

values of the θ/α ratio for an orbital separation of 4° and different locations of the earth station.

It can be seen that for a practical situation where the earth station is not too far away from the wanted satellite, the θ/α ratio remains in the range 1.1 to 1.18. A practical rule of thumb is:

$$\theta = 1.15\alpha \qquad (5.81)$$

Uplink interference analysis. The uplink interference deals with the case where a satellite transponder receives a wanted carrier from an earth station located within the coverage of its receiving antenna and some carrier power at the same frequency from an interfering station normally transmitting to an adjacent satellite, as illustrated in Figure 5.33. The worst case is assumed, i.e. the wanted

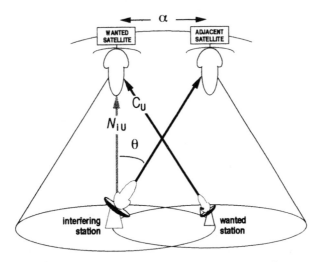

Figure 5.33 Uplink interference from adjacent satellite systems.

earth station is located at the edge of coverage of the satellite it transmits to, and the interfering earth station is located at the centre of coverage.

From (5.59) and (5.60), the expression for carrier-to-interference power ratio $(C/N_i)_U$ is:

$$\left(\frac{C}{N_i}\right)_U = \frac{EIRP_{ESw}}{EIRP_{ESi}} \frac{B_i}{min[B_i, B_N]} \frac{L_{Ui}}{L_{Uw}} \frac{G_{RXw}}{G_{RXi}}$$

$$\left(\frac{C}{N_i}\right)_U dB = EIRP_{ESw}(dBW) - EIRP_{ESi}(dBW) + 10\log B_i - 10\log min[B_i, B_N]$$

$$+ L_{Ui}(dB) - L_{Uw}(dB) + G_{RXw}(dBi) - G_{RXi}(dBi) \qquad (5.82)$$

where:

$EIRP_{ESw}$ is the EIRP for the wanted carrier in the direction of the wanted satellite;

$EIRP_{ESi}$ is the EIRP for the interfering carrier in the direction of the wanted satellite;

L_{Uw} and L_{Ui} are the uplink path loss for the wanted carrier and the interfering one, respectively;

G_{RXw} and G_{RXi} are the satellite composite receive gain values given by (5.61) in the direction of the wanted and interfering earth stations, respectively.

The antenna radiation pattern of the interfering earth station is such that:

$$EIRP_{ESi}(dBW) = EIRP_{ESi,max}(dBW) - G_{Ti,max}(dBi) + 29 - 25\log\theta \qquad (5.83)$$

where $EIRP_{ESi,max}$ is the maximum value of the EIRP allocated to the interfering carrier of bandwidth B_i by the interfering earth station with maximum transmitting antenna gain $G_{Ti,max}$.

Assuming limited geographical area, $L_{Ui}(dB) = L_{Uw}(dB)$. Considering the worst case, where the wanted earth station is at edge of coverage and the interfering one at centre of coverage:

$$G_{RXw}(dBi) = G_{RXi}(dBi) - 3\,dB$$

Putting $\theta = 1.15\alpha$ according to (5.81), then the expression for $(C/N_i)_U$ is:

$$\left(\frac{C}{N_i}\right)_U (dB) = EIRP_{ESw}(dBW) - EIRP_{ESi,max}(dBW) + G_{Ti,max}(dBi) - 32$$
$$+ 25\log(1.15\alpha) + 10\log B_i - 10\log min[B_i, B_N] \qquad (5.84)$$

The corresponding carrier to interference spectral power density ratio $(C/N_{oi})_U$ is given by:

$$\left(\frac{C}{N_{oi}}\right)_U (dBHz) = \left(\frac{C}{N_i}\right)_U + 10\log B_N \qquad (5.85)$$

Example:

Consider a satellite angular separation $\alpha = 4°$. The wanted station is a VSAT (1.2 m antenna, operating at 14 GHz) with an $EIRP_{ESw} = 40\,dBW$. The receiver noise bandwidth is $B_N = 100\,\text{kHz}$. The interfering earth station is a typical EUTELSAT SMS station (5 m antenna) with $G_{\text{Ti,max}} = 55\,\text{dBi}$ and $\text{EIRP}_{\text{ESi,max}} = 70\,\text{dBW}$. The interfering carrier bandwidth is $B_i = 2\,\text{MHz}$. From (5.82), $(C/N_i)_U = 23\,\text{dB}$.

Downlink interference analysis. Figure 5.34 illustrates the worst case for downlink interference.

From (5.59) and (5.60), the expression for carrier-to-interference power ratio $(C/N_i)_D$ is:

$$\left(\frac{C}{N_i}\right)_D = \frac{\text{EIRP}_{\text{SLww}}}{\text{EIRP}_{\text{SLiw}}} \frac{B_i}{\min[B_i, B_N]} \frac{L_{\text{Di}}}{L_{\text{Dw}}} \frac{G_{\text{RXw}}}{G_{\text{RXi}}}$$

$$\left(\frac{C}{N_i}\right)_D (\text{dB}) = \text{EIRP}_{\text{SLww}}(\text{dBW}) - \text{EIRP}_{\text{SLiw}}(\text{dBW}) + 10\log B_i - 10\log\min[B_i, B_N]$$

$$+ L_{\text{Di}}(\text{dB}) - L_{\text{Dw}}(\text{dB}) + G_{\text{RXw}}(\text{dBi}) - G_{\text{RXi}}(\text{dBi}) \qquad (5.86)$$

where:

$\text{EIRP}_{\text{SLww}}$ is the wanted satellite EIRP for the wanted carrier in the direction of the wanted station;

$\text{EIRP}_{\text{SLiw}}$ is the interfering satellite EIRP for the interfering carrier in the direction of the wanted station;

L_{Dw} and L_{Di} are the downlink path loss for the wanted carrier and the interfering one, respectively;

G_{RXw} and G_{RXi} are the wanted station composite receive gain values given by (5.61) in the direction of the wanted and interfering satellites, respectively.

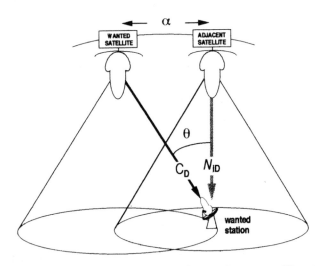

Figure 5.34 Downlink interference from adjacent satellite systems.

For a wanted station at edge of coverage of the wanted satellite and at centre of coverage of the interfering one (worst case):

$$\text{EIRP}_{\text{SLww}}(\text{dBW}) = \text{EIRP}_{\text{SLw,max}}(\text{dBW}) - 3\text{dB}$$

$$\text{EIRP}_{\text{SLiw}}(\text{dBW}) = \text{EIRP}_{\text{SLi,max}}(\text{dBW})$$

Assuming limited geographical area, $L_{\text{Di}}(\text{dB}) = L_{\text{Dw}}(\text{dB})$. The antenna radiation pattern of the wanted station is such that:

$$G_{\text{RXi}}(\text{dBi}) = G_{\text{RXw}}(\text{dBi}) - G_{\text{RXmax}}(\text{dBi}) + 29 - 25 \log \theta \quad (\text{dB}) \qquad (5.87)$$

where $G_{\text{RXmax}}(\text{dBi})$ is the maximum receive composite gain of the wanted station.

Putting $\theta = 1.15\alpha$ according to (5.81), then the expression for $(C/N_i)_{\text{D}}$ is:

$$\left(\frac{C}{N_i}\right)_{\text{D}}(\text{dB}) = \text{EIRP}_{\text{SLw,max}}(\text{dBW}) - \text{EIRP}_{\text{SLi,max}}(\text{dBW}) + 10 \log B_i - 10 \log \min[B_i, B_{\text{N}}]$$

$$+ G_{\text{RXmax}}(\text{dBi}) - 32 + 25 \log(1.15\alpha) \qquad (5.88)$$

The corresponding carrier to interference spectral power density ratio $(C/N_{0i})_{\text{D}}$ is given by:

$$\left(\frac{C}{N_{0i}}\right)_{\text{D}}(\text{dBHz}) = \left(\frac{C}{N_i}\right)_{\text{D}} + 10 \log B_{\text{N}} \qquad (5.89)$$

Example:

Consider a satellite angular separation $\alpha = 4°$. The wanted station is a VSAT (1.2 m antenna, operating at 12 GHz) with $G_{\text{RXmax}} = 41$ dBi, receiving the outbound link from the wanted satellite with $\text{EIRP}_{\text{SLw,max}}(\text{dBW}) = 25$ dBW. The receiver noise bandwidth is $B_{\text{N}} = 500$ kHz. The interfering satellite transmits a TV carrier with $\text{EIRP}_{\text{SLi,max}}(\text{dBW}) = 52$ dBW and bandwidth $= 26$ MHz. From (5.88), $(C/N_i)_{\text{D}} = 15.7$ dB.

5.5.4.2 Terrestrial interference

Bands allocated to satellite communications are most often also allocated to terrestrial microwave links. The only restricted frequency bands are available at Ku-band and Ka-band for exclusive use for satellite communications (see Figure 1.15). In those bands, VSAT networks are protected against terrestrial interference. At C-band, terrestrial interference is a major concern to VSAT networks. Indeed, the smaller the earth station antenna and the lower the frequency, the larger the beamwidth. Therefore small VSATs are sensitive to interference, especially at C-band. VSATs being used for business applications and installed at the user's premises, possibly in city centres, terrestrial interference at C-band may be important. The most convenient way to protect the network from being interfered is to use spread spectrum communications. The calculation of the interference in a city environment is quite complex, and it would be recommended to proceed to site measurements before installing VSATs and hub.

5.5.5 Conclusion

As a result of its small size, the VSAT antenna provides limited interference rejection. Terrestrial interference can be avoided by using exclusive bands at 14/12 GHz and 30/20 GHz. Adjacent satellite interference still remains. Should the adjacent satellites transmit carriers with high EIRP on the frequency of the outbound link, then the VSATs may receive an unacceptable level of interference.

The inbound links, as a result of the low EIRP of the transmitting VSATs, are received at the input of the satellite transponder with low power. Should other stations transmit on the orthogonal polarisation, the uplink part of the inbound link may suffer from cross-polarisation interference, especially if those stations are not properly aligned with the satellite antenna polarisation. The downlink part of the inbound link will also be contaminated by the cross-polarisation interference generated by the carriers conveyed by the cross-polar transponder. It is therefore desirable, when planning a VSAT network, to check the occupancy status of the cross-polar transponder.

Rain or ice induced cross-polar interference may contribute in a sensitive way to the overall level of interference during small percentages of time. This is to be considered when addressing link availability.

5.6 OVERALL LINK PERFORMANCE

The overall link is that from station to station. It comprises two portions : the uplink from the transmitting station to the satellite, and the downlink from the satellite to the receiving station. It is considered here that uplinked carriers are amplified and frequency converted by the satellite transponder before being transmitted on the downlink (no onboard demodulation and remodulation).

The overall carrier power-to-noise power spectral density ratio at the earth station receiver input $(C/N_0)_T$ measures the overall link performance.

Table 5.12 provides a summary of the main equations to be used for evaluating $(C/N_0)_T$. The meaning of each of the notations is given in Table 5.13.

5.7 BIT ERROR RATE DETERMINATION

The overall radio frequency link performance is given by the value of $(C/N_0)_T$ which can be calculated from equatiom (5.7) introducing the parameters discussed in the previous sections.

The discussion will now focus on the ways to derive the quality of the signal delivered to the user terminal, that is the bit error rate (BER).

Table 5.12 Summary of main equations for evaluating the overall link carrier power to noise power spectral density ratio $(C/N_0)_T$

Equation	Unit	Source
Overall link quality		
$(C/N_0)_T^{-1} = (C/N_0)_U^{-1} + (C/N_0)_D^{-1} + (C/N_0)_{IM}^{-1} + \sum(C/N_{0i})_U^{-1} + \sum(C/N_{0i})_D^{-1}$ (all terms in absolute units, not dB)	Hz	(5.7)
Uplink calculations		
$(C/N_0)_U(dBHz) = IBO_1(dB) + (C/N_0)_{Usat}(dBHz)$	dBHz	(5.10)
$IBO_1(dB) = \Phi(dBW/m^2) - \Phi_{sat}(dBW/m^2)$	dB	(5.15)
$\Phi(dBW/m^2) = EIRP_{ES}(dBW) + G_1(dB) - L_U(dB)$	dBW/m²	(5.14)
$(C/N_0)_{Usat}(dBHz) = \Phi_{sat}(dBW/m^2) - G_1(dBi) + (G/T)_{SL}(dBK^{-1}) + 228.6$	dBHz	(5.11)
$G_1(dBi) = 10\log 4\pi + 20\log(f_U/c)$	dBi	(5.12)
Transponder transfer calculations		
$\Phi_t(W/m^2) = \sum\Phi_i(W/m^2)$ $i = 1, 2, \dots, N$ (all terms in absolute units, not dB)	W/m²	
$IBO_t(dB) = \Phi_t(dBW/m^2) - \Phi_{sat}(dBW/m^2)$	dB	(5.16)
$OBO_t(dB) = 0$ if $-5\,dB < IBO_t < 0\,dB$ $0.9\,(IBO_t(dB) + 5)$ if $IBO_t < -5\,dB$		(A6.10)
Downlink calculations		
$(C/N_0)_D(dBHz) = OBO_1(dB) + (C/N_0)_{Dsat}(dBHz)$	dBHz	(5.35)
$OBO_1(dB) = 0.9(IBO_1(dB) + 5)$ if $IBO_t < -5\,dB$ (assumes linear transfer)	dB	(A6.10)
$(C/N_0)_{Dsat}(dBHz) = EIRP_{SLsat}(dBW) - L_D(dB) + (G/T)_{ES}(dBK^{-1}) + 228.6$	dBHz	(5.36)
Intermodulation calculations		
$(C/N_0)_{IM}(dBHz) = 79 - 10\log n - 1.65(IBO_t(dB) + 5)$ if $IBO_t\,dB < -5\,dB$	dBHz	(5.55)
$IBO_t = IBO_1(dB) + 10\log n$ where n is the number of equally powered carriers	dB	(A6.9)
Co-channel interference calculations		
$(C/N_i)_U(dB) = EIRP_{ESw}(dBW) - EIRP_{ESi}(dBW) + 10\log B_i$ $- 10\log\min[B_i, B_N] + L_{Ui}(dB) - L_{Uw}(dB) + G_{RXw}(dBi) - G_{RXi}(dBi)$	dB	(5.62)
$(C/N_{0i})_U(dBHz) = (C/N_i)_U(dB) + 10\log B_N$	dBHz	(5.69)
$(C/N_i)_D(dB) = EIRP_{SL1ww}(dBW) - EIRP_{SL2iw}(dBW) + 10\log B_i - 10\log \min[B_i, B_N]$	dB	(5.63)
$(C/N_{0i})_D(dBHz) = (C/N_i)_D(dB) + 10\log B_N$	dBHz	(5.69)
Cross polarisation interference		
$(C/N_i)_U = \Delta_U\{10^{-(XPI_{TX}/10)} + 10^{-(XPD/10)} + 10^{-(XPI_{RX}/10)}\}^{-1}$	dB	(5.75)
$(C/N_{0i})_D(dBHz) = (C/N_i)_D(dB) + 10\log B_N$	dBHz	(5.77)
$(C/N_i)_D = \Delta_D\{10^{-(XPI_{TX}/10)} + 10^{-(XPD/10)} + 10^{-(XPI_{RX}/10)}\}^{-1}$	dB	(5.75)
$(C/N_{0i})_D(dBHz) = (C/N_i)_D(dB) + 10\log B_N$	dBHz	(5.77)

Table 5.12   (*Contd.*)

Equation	Unit	Source
Adjacent satellite interference		
$(C/N_i)_U(dB) = EIRP_{ESw}(dBW) - EIRP_{ESi,max}(dBW) + G_{Ti,max}(dBi)$ $\quad - 32 + 25\log(1.15\alpha) + 10\log B_i - 10\log\min[B_i, B_N]$	dB	(5.84)
$(C/N_{0i})_U(dBHz) = (C/N_0)_U(dB) + 10\log B_N$	C/N_0	(5.85)
$(C/N_i)_D(dB) = EIRP_{SLw,max}(dBW) - EIRP_{SLi,max}(dBW) + 10\log B_i$ $\quad - 10\log\min[B_i, B_N] + G_{RXmax}(dBi) - 32 + 25\log(1.15\alpha)$	dB	(5.88)
$(C/N_{0i})_D(dBHz) = (C/N_i)_D + 10\log B_N$	dBHz	(5.89)

Table 5.13   Meaning of symbols displayed in Table 5.12

Symbol	Meaning	Unit	Source equation
B_N	Receiver noise equivalent bandwidth	Hz	(5.57)
B_i	Interfering carrier bandwidth	Hz	(5.58)
$(C/N_0)_D$	Downlink carrier power to noise power spectral density ratio	Hz	(5.35)
$(C/N_0)_{Dsat}$	Same as above, at saturation	Hz	(5.36)
$(C/N_{0i})_D$	Downlink carrier power to interference noise power spectral density ratio (Hz)	Hz	(5.89)
$(C/N_0)_{IM}$	Carrier power to intermodulation power spectral density ratio	Hz	(5.55)
$(C/N_{0i})_U$	Uplink carrier power to interference noise power spectral density ratio	Hz	(5.85)
$(C/N_0)_T$	Overall link (from station to station) carrier power to total noise power spectral density ratio	Hz	(5.7)
$(C/N_0)_U$	Uplink carrier power to noise power spectral density ratio	Hz	(5.10)
$(C/N_0)_{Usat}$	Same as above, at saturation	Hz	(5.11)
$EIRP_{ES}$	EIRP of earth station	W	(5.17)
$EIRP_{ESi}$	EIRP of interfering earth station	W	(5.17)
$EIRP_{ESw}$	EIRP of wanted earth station	W	(5.17)
$EIRP_{SLsat}$	EIRP of satellite transponder at saturation	W	(5.37)
$EIRP_{SL1ww}$	EIRP of satellite in beam 1 for wanted carrier in direction of wanted station	W	(5.37)
$EIRP_{SL2iw}$	EIRP of satellite in beam 2 for interfering carrier in direction of wanted station	W	(5.37)
G_{RXi}	Satellite composite receive gain of beam 1 for interfering carrier		(5.67)
G_{RXw}	Satellite composite receive gain of beam 1 for wanted carrier		(5.62)

(*continued*)

Table 5.13   *(Contd.)*

Symbol	Meaning	Unit	Source equation
G_1	Gain of an ideal antenna with area equal to 1 m²		(5.12)
$(G/T)_{ES}$	Figure of merit of earth station receiving equipment	K^{-1}	(5.42)
$(G/T)_{SL}$	Figure of merit of satellite receiving equipment	K^{-1}	(5.31)
IBO_t	Total input back-off		(5.16)
IBO_1	Input back-off per carrier		(5.15)
L_U	Uplink pathloss		(5.27)
L_{Ui}	Uplink path loss for interfering carrier		(5.27)
L_{Uw}	Uplink path loss for wanted carrier		(5.27)
L_D	Downlink path loss		(5.41)
OBO_t	Total output back-off per carrier with multicarrier operation mode		(A6.6)
OBO_1	Output back-off per carrier with multicarrier operation mode		(A6.10)
XPD	Cross-polar discrimination		(A4.6)
XPI_{RX}	Receive antenna cross polarisation isolation		(A4.5)
XPI_{TX}	Transmit antenna cross polarisation isolation		(A4.5)
Δ	Ratio of co-polar wanted carrier power to cross-polar inter-fering carrier power		(5.75)
Φ_{sat}	Power flux density at saturation (W/m²) for satellite transponder	W/m²	(5.11) (5.14)

The BER is a function of the dimensionless ratio E_b/N_0 where E_b is the energy per information bit and N_0 the overall link noise power spectral density, i.e. $N_0 = N_{0T}$ as expressed by equation (5.5). The energy per information bit is defined as the energy accumulated at the receiver from reception of carrier power C_D during a time interval equal to the time it takes to receive an information bit. At information bit rate R_b, the time it takes to receive an information bit is $T_b = 1/R_b$. Therefore:

$$E_b = \frac{C_D}{R_b} \quad (J)$$

(5.90)

and

$$\frac{E_b}{N_0} = \frac{C_D/R_b}{N_{0T}} = \frac{(C/N_0)_T}{R_b}$$

$$\frac{E_b}{N_0}(dB) = \left(\frac{C}{N_0}\right)_T (dBHz) - 10\log R_b$$

(5.91)

The relationship between E_b/N_0 and BER depends on the type of modulation and forward error correction (FEC) scheme used. FEC techniques are based on transmitting redundant bits along with the information bits, in such a way as to allow the receiver to decode and correct some of the erroneous bits. These techniques are presented in [MAR93, pp. 114–117]. It must be made clear that, contrary to ARQ protocols presented in Chapter 4, Section 4.5, FEC does not guarantee error-free transmission. FEC only reduces the BER, given $(C/N_0)_T$, at the expense of a larger bandwidth.

Figure 5.35 displays a comparison between an uncoded link and a coded one using BPSK or QPSK modulation. Curve 1 represents the uncoded link perform-ance and curves 2 the coded ones. It is assumed here that both links deliver data to the user at same rate R_b and same BER. The coding gain G_{cod} is defined as the E_b/N_0

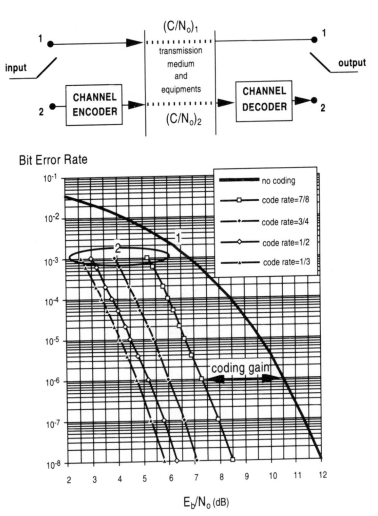

Figure 5.35 Comparison between uncoded and coded link for convolutional encod-ing and typical Viterbi decoder performance.

saving introduced by the coding scheme at the considered BER, as illustrated in Figure 5.35 for a BER of 10^{-6} and a code rate $\rho = 7/8$:

$$G_{cod}(dB) = \left(\frac{E_b}{N_0}\right)_1 (dB) - \left(\frac{E_b}{N_0}\right)_2 (dB) \tag{5.92}$$

This saving translates into a reduction in the required $(C/N_0)_T$:

No coding: $$\left(\frac{C}{N_0}\right)_{T1} (dBHz) = \left(\frac{E_b}{N_0}\right)_1 (dB) + 10\log R_b \tag{5.93}$$

With coding: $$\left(\frac{C}{N_0}\right)_{T2} (dBHz) = \left(\frac{E_b}{N_0}\right)_2 (dB) + 10\log R_b \tag{5.94}$$

The reduction in $(C/N_0)_T$ is equal to:

$$\Delta\left(\frac{C}{N_0}\right)_T (dB) = \left(\frac{C}{N_0}\right)_{T1} (dBHz) - \left(\frac{C}{N_0}\right)_{T2} (dBHz)$$

$$= \left(\frac{E_b}{N_0}\right)_1 (dB) - \left(\frac{F_b}{N_0}\right)_2 (dB)$$

$$= G_{cod} \tag{5.95}$$

The coding gain is higher with low values of code rate.

5.8 POWER VERSUS BANDWIDTH EXCHANGE

For a carrier conveying an information bit rate R_b, using forward error correction (FEC) with code rate ρ, and modulation with spectral efficiency Γ (ratio of bit rate to bandwidth), the required bandwidth is [MAR93, p. 115]:

$$B = \frac{R_b}{\rho\Gamma} \tag{5.96}$$

Hence, a larger bandwidth is required for low values of code rate, given the information bit rate R_b and the modulation scheme. However, with low code rate one benefits from a larger coding gain and therefore a lower requirement on E_b/N_0 which translates, according to equation (5.95), into a reduced requirement for the received carrier power. This suggests a potential power versus bandwidth exchange, which will now be demonstrated through a specific example.

Consider a VSAT transmitting at an information bit rate $R_b = 64\,\text{kb/s}$. BPSK modulation is used with spectral efficiency $\Gamma = 0.7\,\text{b/s/Hz}$. The required bit error rate is BER $= 10^{-7}$. Table 5.14 gives the required values of E_b/N_0 depending on the code rate, as given by Figure 5.35, the corresponding values of $(C/N_0)_T$, according to formulas (5.93) and (5.94), and bandwidth as given by formula (5.96).

Figure 5.36 illustrates the power versus bandwidth exchange, which is experienced by changing the code rate. This flexible exchange is paramount to any design of a VSAT network.

Example 221

Table 5.14 Impact of code rate on required values of radio frequency link parameters (BPSK, $R_b = 64$ kb/s, BER $= 10^{-7}$)

Code rate	Required E_b/N_0	Required $(C/N_0)_T$	Required bandwidth
1	11.3 dB	59.4 dBHz	92 kHz
7/8	8 dB	56.1 dBHz	104 kHz
3/4	6.6 dB	54.7 dBHz	122 kHz
2/3	6.2 dB	54.3 dBHz	137 kHz
1/2	5.8 dB	54 dBHz	183 kHz

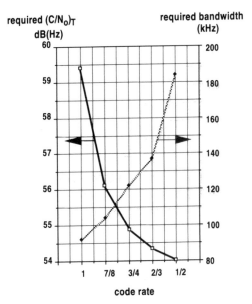

Figure 5.36 Illustration of the power versus bandwidth exchange for $R_b = 64$ kb/s and BER $= 10^{-7}$.

5.9. EXAMPLE

The purpose of this section is to perform a typical dimensioning of radio frequency links according to the principles exposed in this chapter.

Figure 5.37 shows the architecture of the considered VSAT network.

Uplink and downlink frequencies are all approximated respectively to $f_U = 14.25$ GHz and $f_D = 12.7$ GHz. The service requirement is a BER $= 10^{-7}$, which entails a required $E_b/N_0 = 11.3$ dB without coding. The earth stations are assumed to be located so that the minimum elevation angle for all stations is $E_{min} = 35°$. For

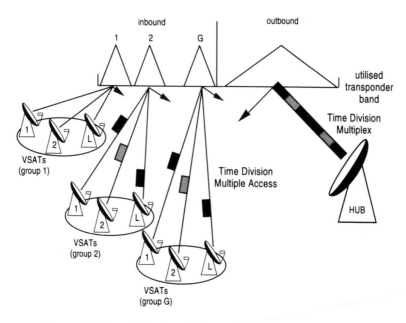

Figure 5.37   architecture of the considered VSAT network.

Table 5.15   Network parameters

1	Type of access		FDMA/TDMA
2	Number of groups of VSATs	G	3
3	Number of VSATs per groups	L	33
4	Number of VSATs in network	N	99
5	Inbound information bit rate	R_{binb}	64 kb/s
6	Inbound code rate	ρ_{inb}	0.5
7	Inbound transmission rate	R_{cinb}	128 kb/s
8	Inbound modulation	BPSK	
9	Inbound modulation spectral efficiency	Γ	0.7 b/s. Hz
10	Outbound transmission bit rate	R_{binb}	256 kb/s
11	Outbound code rate	ρ_{outb}	0.5
12	Outbound transmission bit rate	R_{coutb}	512 kb/s
13	Outbound modulation	BPSK	
14	Outbound modulation spectral efficiency	Γ	0.7 b/s.Hz
15	Percentage for guard bands		20%
16	Utilised transponder band		1536 kHz

Notes:
Line 4 = Line 2 × Line 3
Line 7 = Line 5/Line 6
Line 12 = Line 10/Line 11
Line 16 = (1 + Line 15/100) × (Line 2 × Line 7/Line 9 + Line 12/Line 14)
according to (5.97)

Example 223

instance, the earth station can be at medium latitude (about 45°) and within 20° of relative longitude with respect to the satellite.

Table 5.15 displays the network parameters; Table 5.16 gives the satellite, VSAT and hub characteristics; Table 5.17 indicates link parameter calculations for clear

Table 5.16 VSAT network satellite and earth station parameters

	Reference	Symbol	Value	Unit
Satellite (SL) parameters				
Saturation power flux density	(5.11)	Φ_{sat}	-85	dBW/m²
Edge of coverage saturation EIRP	(5.37)	$EIRP_{SLsat}$	43	dBW
Edge of coverage satellite G/T	(5.31)	$(G/T)_{SL}$	2.5	dBK^{-1}
Transponder bandwidth		B_{Xpond}	36	MHz
VSAT parameters				
Antenna diameter	App. 4	D_{VSAT}	1.8	m
HPA rated power		$P_{TXVSATsat}$	1	W
Transmit gain (max.)	(5.23)	$G_{TmaxVSAT}$	46.4	dB
TX implementation losses:				
feeder loss	§ 5.2.2	$L_{FTX, VSAT}$	0.2	dB
depointing loss	(5.24)	$L_{TmaxVSAT}$	1.2	dB
HPA output back-off	(A6.1)	OBO_{VSAT}	-6	dB
HPA output power	(A6.1)	P_{TXVSAT}	0.25	W
EIRP	(5.17)	$EIRP_{VSAT}$	39	dBW
Receive gain (max.)	(5.44)	$G_{RmaxVSAT}$	45.4	dBi
RX implementation losses:				
feeder loss	§5.3.4	$L_{FRX, VSAT}$	0.5	dB
depointing loss	(5.46)	$L_{RmaxVSAT}$	0.9	dB
Figure of merit (clear sky)	§5.3.4	$(G/T)_{VSAT}$	22.3	dBK^{-1}
Figure of merit (rain)	§5.3.4		18.8	dBK^{-1}
Hub parameters				
Antenna diameter	App. 4	D_{HUB}	5.5	m
HPA rated power		P_{TXHUB}	5	W
Transmit gain (max.)	(5.23)	$G_{TmaxHUB}$	56.3	dBi
TX implementation losses:				
feeder loss	§5.2.2	$L_{FTX, HUB}$	0.2	dB
depointing loss (tracking)	§5.2.2	$L_{TmaxHUB}$	0.5	dB
HPA output back-off	(A6.1)	OBO_{HUB}	-8	dB
HPA output power	(A6.1)	P_{TXHUB}	0.8	W
EIRP	(5.17)	$EIRP_{HUB}$	54.6	dBW
Receive gain (max.)	(5.44)	$G_{RmaxHUB}$	55.1	dBi
RX implementation losses:				
feeder loss	§5.3.4	$L_{FRX, VSAT}$	0.5	dB
depointing loss (tracking)	§5.2.2	$L_{RmaxVSAT}$	0.5	dB
Figure of merit (clear sky)	§5.3.4	$(G/T)_{HUB}$	32.5	dBK^{-1}
Figure of merit (rain)	§5.3.4	$(G/T)_{HUB}$	29	dBK^{-1}

sky and rain conditions. (The selected value of rain attenuation, $A_{RAIN} = 6\,dB$, is exceeded for 0.01% of the time in Europe climatic region H, at an elevation angle $E = 35°$.)

The network comprises $G = 3$ groups of $L = 33$ VSATs each, so that the total number of VSATs in the network is 99. Each group is being allocated a frequency band for its inbound links. The VSATs of a group access this frequency band in a Reservation/Random TDMA mode. Therefore the inbound links are made of bursts of inbound carriers. These bursts convey information bits at a rate $R_{binb} = 64\,kb/s$. Coding with code rate $\rho_{inb} = 1/2$ is used, with coding gain $= 5.5\,dB$ at the required BER $= 10^{-7}$. Therefore the inbound transmitted bit rate within each burst is $R_{cinb} = R_{binb}/\rho_{rinb} = 128\,kb/s$. The modulation is BPSK, with spectral efficiency $\Gamma = 0.7\,b/s.Hz$. Therefore, according to (5.96) the utilised bandwidth per inbound link is $B_{inb} = 183\,kHz$.

The outbound link is a time division multiplex (TDM) at information bit rate $R_{boutb} = 256\,kb/s$ with code rate $\rho_{outb} = 1/2$. The transmitted bit rate is $R_{coutb} = 512\,kb/s$. The modulation is BPSK. According to (5.96), the utilised bandwidth for the outbound link is $B_{outb} = 732\,kHz$.

Assuming 20% guard band between carriers, the total utilised transponder bandwidth is:

$$B = 1.2(GB_{inb} + B_{outb}) = 1.2 \times (3 \times 183 + 732)\,kHz = 1.54\,MHz \qquad (5.97)$$

The percentage of transponder bandwidth utilised by the VSAT network is:

$$100\frac{B}{B_{Xpond}} = 100\frac{1.54}{36} = 4.3\% \qquad (5.98)$$

Table 5.17   Link parameter calculations

	Inbound		Outbound	
	Clear sky	Rain on VSAT uplink	Clear sky	Rain on VSAT downlink
Uplink				
1 Earth station EIRP (per carrier)	39 dBW	39 dBW	54.6 dBW	54.6 dBW
2 Path loss at 35° elevation angle	208 dB	214 dB	208 dB	208 dB
3 Gain of ideal 1 m² antenna	44.5 dBi	44.5 dBi	44.5 dBi	44.5 dBi
4 Operating flux density (per carrier)	−124.5 dBW/m²	−130.5 dBW/m²	−108.8 dBW/m²	−108.8 dBW/m²
5 Saturation flux density	−85 dBW/m²	−85 dBW/m²	−85 dBW/m²	−85 dBW/m²

Example 225

Table 5.17 (*Contd.*)

		Inbound		Outbound	
		Clear sky	Rain on VSAT uplink	Clear sky	Rain on VSAT downlink
6	Input back-off (per carrier)	−39.5 dB	−45.5 dB	−23.8 dB	−23.8 dB
7	Total input back-off	−23.5 dB	−23.6 dB	−23.5 dB	−23.6 dB
8	Satellite G/T	$2.5\,\mathrm{dBK}^{-1}$	$2.5\,\mathrm{dBK}^{-1}$	$2.5\,\mathrm{dBK}^{-1}$	$2.5\,\mathrm{dBK}^{-1}$
9	Uplink C/N_0 at saturation	101.6 dBHz	101.6 dBHz	101.6 dBHz	101.6 dBHz
10	Uplink C/N_0 (per carrier)	62.1 dBHz	56.1 dBHz	77.7 dBHz	77.7 dBHz
	Downlink				
11	Saturated satellite EIRP	43 dBW	43 dBW	43 dBW	43 dBW
12	Path loss at 35° elevation angle	207 dB	207 dB	207 dB	213 dB
13	Earth station G/T	$32.5\,\mathrm{dBK}^{-1}$	$32.5\,\mathrm{dBK}^{-1}$	$22.3\,\mathrm{dBK}^{-1}$	$18.8\,\mathrm{dBK}^{-1}$
14	Downlink C/N_0 saturation	97.1 dBHz	97.1 dBHz	86.9 dBHz	77.4 dBHz
15	Total output back-off	−16.6 dB	−16.7 dB	−16.6 dB	−16.7 dB
16	Output back-off (per carrier)	−31.0 dB	−36.4 dB	−17.0 dB	−17.0 dB
17	Small signal suppression	0.5 dB	0.5 dB	0 dB	0 dB
18	Operating EIRP (per carrier)	11.5 dBW	6.1 dBW	26.0 dBW	26.0 dBW
19	Downlink C/N_0 (per carrier)	65.5 dBHz	60.1 dBHz	70.0 dBHZ	60.5 dBHz
20	*Intermodulation*	104.7 dBHz	104.9 dBHz	119.3 dBHz	124.9 dBHz
21	*Interference*	72.8 dBHz	67.4 dBHz	77.6 dBHz	77.6 dBHz
22	*Total link C/N_0 (per carrier)*	60.2 dBHz	54.4 dBHz	68.7 dBHz	60.3 dBHz
23	Required E_b/N_0 (no coding)	11.3 dB	11.3 dB	11.3 dB	11.3 dB

(*continued*)

Table 5.17 *(Contd.)*

		Inbound		Outbound	
		Clear sky	Rain on VSAT uplink	Clear sky	Rain on VSAT downlink
24	Coding gain	5.5 dB	5.5 dB	5.5 dB	5.5 dB
25	Required E_b/N_0 (with coding)	5.8 dB	5.8 dB	5.8 dB	5.8 dB
26	Required C/N_0	53.9 dBHz	53.9 dBHz	59.9 dBHz	59.9 dBHz
27	Margin	6.4 dB	0.6 dB	8.8 dB	0.4 dB
28	Transponder power usage per carrier	0.08%	0.02%	2.02%	2.02%
29	Transponder power usage	2.16%	2.13%	2.16%	2.13%
30	Transponder bandwidth usage per carrier	0.51%	0.51%	2.03%	2.03%
31	Transponder bandwidth usage	4.27%	4.27%	4.27%	4.27%

Notes:

Line 2: equation (5.27) $L_{FS} = 206.6$ dB (Figure 5.13) $+ 0.9$ dB (Figure 5.14)
$L_A = 0.5$ dB $+ A_{RAIN}$; clear sky: $A_{RAIN} = 0$ dB, rain: $A_{RAIN} = 6$ dB
Line 4: equation (5.14) = Line 1 − Line 2 + Line 3
Line 6: equation (5.15) = Line 4 − Line 5
Line 7: equation (5.16) $= 10 \log (3 \times 10^{IBO_{inb}(dB)/10} + 10^{IBO_{outb}(dB)/10})$ if clear sky
$= 10 \log (2 \times 10^{IBO_{inb}(dB)/10} + 10^{IBO_{inbrain}(dB)/10} + 10^{IBO_{outbrain}/10})$ if rain
Line 9: equation (5.11) = Line 5 − Line 3 + Line 8 + 228.6
Line 10: equation (5.10) = Line 9 + Line 6
Line 12: equation (5.41) $= L_{FS} = 205.6$ dB (Figure 5.13) $+ 0.9$ dB (Figure 5.14)
$L_A = 0.5$ dB $+ A_{RAIN}$; clear sky: $A_{RAIN} = 0$ dB, rain: $A_{RAIN} = 6$ dB
Line 14: equation (5.36) = Line 11 − Line 12 + Line 13 + 228.6
Line 15: equation (A6.10) = 0.9 (Line 7 + 5)
Line 16: equation (A6.10) = 0.9 (Line 6 + 5)
Line 17: small signal suppression takes into account the capture effect (see section 5.1.4)
Line 18: Line 11 + Line 16 − Line 17
Line 19: equation (5.35) = Line 14 + Line 16 − Line 17
Line 20: equation (5.55) $= 79 − 10 \log 3 − 1.65$ (Line 7 + 5) for inbound links
$=$ previous result $−$ $EIRP_{inb} + EIRP_{outb}$, for outbound links
Line 21: considers downlink adjacent satellite interference only, see equation (5.88), with $\alpha = 4°$,
$EIRP_{SLi,max} = 52$ dBW, and $B_i = 26$ MHz
$=$ Line 18 $− 52 + 10 \log B_i + G_{RXmax} − 32 + 25 \log 4.6°$ where $G_{RXmax} = 54.6$ dBi inbound,
44.9 dBi outbound
Line 22: equation (5.7) $= − 10 \log (10^{-Line10/10} + 10^{-Line19/10} + 10^{-Line20/10} + 10^{-Line21/10})$
Line 25 = Line 23 − Line 24
Line 26 = Line 25 + $10 \log R_b$ where $R_b = 64$ kb/s inbound, 256 kb/s outbound
Line 27 = Line 22 − Line 26
Line 28 $= 100 \times 10^{OBO_{inb}/10}$ or $100 \times 10^{OBO_{outb}/10}$
Line 29 $= 100 \times 10^{OBO_{tot}/10}$
Line 30 $= 100 B_N/B_{Xpond}$
Line 31 $= 100((\text{Line 16 of Table 5.15})/B_{Xpond})$ according to equation (5.98)

6 FUTURE DEVELOPMENTS

The aim of this chapter is to discuss briefly possible future developments of VSAT technology. The introduction of this book emphasised the fact that VSAT networks are part of the historical evolution of earth station technology towards a reduction in size and cost. From the link analysis developed in Chapter 5, it has become clear that this evolution was possible thanks to the availability of satellite transponders with high EIRP and G/T. Such satellites emerged as a result of coverage design, evolving from global to multiple spot beams, and eventually adaptation to offering regional services, as mentioned in Chapter 2, section 2.1.3.

6.1 COST REDUCTION OF VSATs

Offering small sized stations at low cost does not necessarily mean that VSAT networks are guaranteed a market. Indeed, VSAT networks face terrestrial competition and have developed in the past, and will continue to do so, only if they offer services to the customer at a lower cost than terrestrial alternatives. It is therefore important that VSATs remain cheap and reliable. This will most certainly prevent any drastic technology change which would not offer a sizeable cost reduction.

Reducing the antenna size reduces its cost but increases the risk of interference, unless higher frequency bands are used. However, when changing from Ku-band to Ka-band, three difficulties arise: the lack of satellites operating at Ka-band, the increased cost of the earth station equipment, and the larger link margins to keep up with the same link availability.

Therefore, the most probable evolution concerns the earth station electronic equipment. What is likely is a progressive replacement of the analogue parts of the receiver and transmitter by digital circuits for those functions that can be based on digital processing of signals, such as demodulation and modulation. This would allow more flexibility, as the operational parameters of a VSAT could be reconfigured by software. One single VSAT model could be able to cope with a large variety of applications, from data to compressed video. This would pave the way to a larger mass production of VSAT units, with a subsequent cost reduction.

6.2   NEW SERVICES

Besides being cost effective, VSAT networks also have to satisfy the customer's needs. Today, business traffic is mostly telephony and data, with the low rate data component being presently handled conveniently by VSAT networks. The traditional environment of a host computer communicating with multiple workstations has been progressively phased out in favour of peer-to-peer processing based on local area networks (LANs). With the installation of LANs at the different sites, there has been an increasing demand for interconnecting distant LANs. By offering digital connections to all network sites with bandwidth supplied to each site upon demand, VSAT networks provide an appropriate solution for typically bursty LAN traffic, while retaining the ability to broadcast the same information, such as databases or applications updates, to a number of remote sites. In the future, business traffic will increasingly incorporate other components, in line with the foreseen development of multimedia and mobile services [HAR92].

6.2.1   LAN interconnection

In Europe, the European Commission COST 226 project, with strong support from the European Space Agency, has been studying the feasibility of interconnecting existing LANs of different types and their applications, by means of high speed (typically 2 Mb/s) VSATs. The final report is planned for 1995, and equipment development should follow on rapidly.

Typical data rates on LANs are 4, 10 and 16 Mb/s, much higher than rates provided on VSAT links. However, the majority of the traffic on a LAN is intended for another device on that same LAN. Traffic between LANs typically represents only 5% to 15% of the total traffic on the LAN. Therefore, the data rate required on the satellite link is far below that required within the LAN itself, and can be accommodated by VSAT technology [CAC91], considering that this technology could at reasonable expense convey data rates up to about 2 Mb/s.

The difficulty in interconnecting LANs with VSAT networks resides in the higher rate than that of conventional VSATs and also the specific inter-LAN traffic characteristics, which has been identified as been 'self-similar' (or fractal) [LEL91] [LEL93]. The access protocols described in Chapter 4, which combine random access with some form of reservation, are not suitable at high data rates because they require usage of large transponder bandwidth which is expensive. For instance, information bit rates of 2 Mb/s may entail using a transponder bandwidth of, say, 8 to 10 MHz if coding is considered, along with a normalised throughput in the range from 10 to 60%. Therefore, some type of efficient demand based control access has to be considered instead. A protocol recently implemented on a commercial system is the Advanced Business Communications via Satellite (ABCS) protocol [DOR92]. Meanwhile, research in the field is active, with on-going evaluation of protocols such as the FIFO Ordered Demand Assignment (FODA) protocol [CEL91], the Combined Fixed/Reservation Assignment (CFRA) protocol [ZEI95], and the Combined Free/Demand Assignment Multiple Access (CFDMA) protocol [LEN93].

The LAN/VSAT interface must support all the end-to-end functions required by the LANs and their applications. For instance, the LAN/VSAT interface must have the ability to automatically learn addresses on the locally attached LANs, filter traffic on the LAN and only forward over the satellite link those messages destined for a terminal residing at a distant LAN, attached either to the hub or to another VSAT. The VSAT network performs routeing of messages from one VSAT to another, enabling communication between LANs. The hub acts as a central switch, switching traffic received from one remote LAN to the LAN on which the addressed terminal resides. Such routeing implies address mapping, addresses in the VSAT network being independent from the addresses of the end terminal within its host LAN.

The LAN/VSAT interface must also support end-to-end functions at the transport level. Connection-oriented transport protocols are responsible for error recovery, flow control and packet resequencing on the connection between the sender and the receiver. It has been shown in Chapter 4 that the performance of error control mechanisms depends upon the product $R_b \times T_{RT}$ of the transfer rate R_b and the round trip time (T_{RT}). This product measures the amount of data that would fill the connection, and indicates the buffer space required at the sender and receiver to obtain the maximum throughput on the connection. It represents the amount of unacknowledged data the protocol must handle in order to maintain the connection full. Protocol performance problems arise when the $R_b \times T_{RT}$ product is large, and such a connection is referred to as a 'long fat connection'. A network supporting such a connection is referred to as an LFN, pronounced 'elephan(t)' [RES93, p. 116].

Many applications in the workstation area and most of the applications using the Internet use the Transmission Control Protocol/Internet Protocol (TCP/IP) as the underlying transport protocol. A simple calculation will show the influence of the $R_b \times T_{RT}$ product on the throughput. The considered network architecture is that of Figure 6.1 [YAN92].

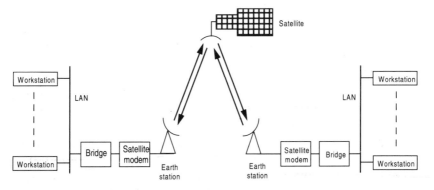

Figure 6.1 Satellite interconnected LANs. (Reproduced from [YAN92] by permission of the Institute of Electrical and Electronics Engineers, Inc., © 1992 IEEE.)

Workstations associated with two distantly separated LANs are interconnected using two wide area bridges. The workstations support the TCP/IP protocol. The LANs are ANSI/IEEE 802.3 Ethernet type operating at 10 Mb/s. Two media access control (MAC) layer bridges are used as LAN interconnection devices. The TCP/IP protocol uses a 16 bit field to report the receive window size to the sender. Therefore, the largest window size that can be used is $2^{16} = 64$ kbytes. For the VSAT network, assuming LAN interconnection between a VSAT and the hub in a star shaped network, or between VSATs in a mesh network, the round trip time is about two times the hop delay, so $T_{RT} \approx 2 \times 270$ ms $= 540$ ms. The theoretical maximum throughput is given by:

$$\text{Throughput} = \text{window size}/\text{round trip time}$$

$$= (64 \text{ kbytes} \times 8 \text{ bits/byte})/0.54 = 948 \text{ kb/s}.$$

It is not possible to get more from the connection, no matter how wide the channel capacity really is. Taking into account the various headers that are introduced by the encapsulation from the TCP layer to the physical layer, the actual maximum throughput would be even lower. Interconnecting LANs by VSATs in a star shaped network would prove even worse, as the delay would be twice as large. To solve the problem, some TCP implementations provide for a mechanism called window scaling, which allows window sizes greater than 64 kbytes.

It can be shown that the bit error rate also impacts on the TCP/IP protocol performance, and a satellite link should provide a BER of less than 10^{-7} to avoid severe degradation [YAN92]. Finally, the slow start mechanism which is used in TCP/IP for solving congestion control problems is highly inefficient over satellite links.

Therefore, new versions of TCP are necessary, or new transport protocols. There is ample matter here for discussion between satellite communications people and the computer science community.

6.2.2 Multimedia

As of now, most existing LANs are not suitable for the transmission of delay sensitive traffic such as voice or video, so separate local networks are needed to support these. Figure 6.2 illustrates how VSAT networks are able to interconnect both types of networks.

In the future, the end user will ask for a combination of services including text, graphics, video, audio, and possibly animation, on a desktop personal computer, and the separate local networks for data and voice attached to the VSATs of Figure 6.2 will merge into a single one.

In any case, candidate satellite link protocols should be able to simultaneously support two types of traffic [BAU91]:

—*stream traffic*, typical for exchange of information such as voice and video processed in real time between end systems producing a continuous flow of

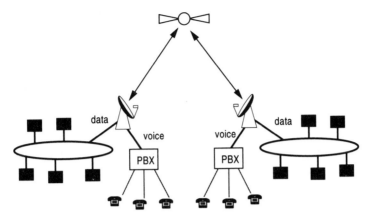

Figure 6.2 Interconnection of voice and data networks using a VSAT network. PBX: Private Branch Exchange.

bits with low tolerance for delay, jitter, etc., therefore usually supported by circuit-switched networks;

—*bursty traffic*, typical for exchange of digital data between computer based end systems, storing or processing those data not in real time, and best suited for packet oriented technologies.

Each service component generates its own type of traffic, with characteristics that largely differ from one component to another in terms of traffic volume, burstiness, error rate, transmission delay and delay jitter. This leads to different and possibly conflicting requirements concerning the VSAT network design, and delicate optimisation procedures remain to be explored.

6.2.3 Mobile services

The trend towards smaller VSAT earth stations, when used at Ka-band, may lead to the emergence of terminals capable of offering the user a higher level of sophistication and a larger set of functionalities than those offered by the foreseen more simple portable or handset terminals of mobile satellite systems. This can be viewed as a form of 'enhanced personal communications services', at the boundary of mobile and fixed communication services. Indeed, such terminals would make available the ISDN services to an increasingly mobile population. Examples of such terminals could be [GAR93]:

—*briefcase portable office*: such a terminal would offer the functionality of a laptop multimedia terminal with facilities equivalent to those provided by basic rate ISDN communications. Therefore, the user would be provided with possible simultaneous voice, data and low rate video connections. Referring to the well established architecture of a traditional VSAT station, one could imagine the 'outdoor' unit as a briefcase hosting the 'indoor' unit, in the form of a standard multimedia laptop computer. Connections between the laptop and the satellite terminal could be wired as with today's VSATs, or more conveniently wireless, with the laptop possibly functioning independently of its associated briefcase while being plugged into terrestrial ISDN access where available;

—*home office terminal*: considering the change in work habits which is leading to an increasing proportion of people installing their office at home or away from urban locations as a result of improved communications and timesharing of work, the home office terminal should offer ISDN capabilities up to 2 Mb/s. This could be supported by a 30–40 cm Ka-band antenna in those areas not yet served by cable or fibre facilities. Such an antenna would be easily installed on a small house or cottage.

6.3   VSATs AND ONBOARD PROCESSING SATELLITES

The satellite payload of Figure 2.4 offers the two functions, switching and multiplexing, provided by an earth hub station when such a hub station is used to support interactivity between remote sites. Therefore, it is conceivable to implement a hubless VSAT network using an onboard processing satellite as a hub in the sky. This would combine the advantages of a hub on the ground (reduced transmit bit rate for remote VSATs) with the lower delay of a fully meshed VSAT network. But this entails onboard demodulation of uplinked carriers, processing of data, and remodulation of carriers.

Such a concept is not nearby: indeed, there is no indication that the design of the planned satellites in the next few years will incorporate onboard processing for commercial use. The satellites designed now will be operating, once launched, for 15 to 20 years, and will offer transponder capacity for VSAT networking as of today. Moreover, the onboard processing places rigid constraints on the earth station requirements in terms of carrier format, bit rates, etc., as can be seen from today's experimental launched systems (ACTS and ITALSAT). It is very unlikely that will emerge in the near future a world-wide market for a unique type of VSAT networks served by dedicated onboard processing satellites.

6.4   USE OF NON-GEOSTATIONARY SATELLITES

Up till now, only geostationary satellites have been considered. World-wide satellite communication systems based on non-geostationary satellites are announced for the end of this decade, such as Motorola's IRIDIUM, Loral's GLOBALSTAR, and others [MAR94]. Although these systems are presently planned for low data rate mobile services, of the order of 10 kb/s, they may be suited for some VSAT applications. Later generation systems, such as the Calling network, now named Teledesic, will offer larger data rates [TUC94]. Actually, the system is

Table 6.1 Evolution trends for VSAT networks (from most likely: $++$ to most unlikely: $--$)

Type of evolution	Time scale	
	Short term (within 5 years)	Long term (within 10 years)
Digital processing of signals at VSAT	$++$	$++$
Cost reduction of station equipment	$++$	$++$
Use of Ka-band	$--$	$+$
New services:		
LAN interconnection	$++$	$++$
multimedia traffic	$+$	$++$
mobile users	$-$	$++$
Satellite onboard processing	$--$	$--$
Use of non-geostationary satellites	$--$	$++$
High quality network management	$++$	$++$

planned to service small fixed earth stations. It may provide interesting alternatives to geostationary satellite based VSAT networks.

6.5 NETWORK MANAGEMENT

VSAT networks, as all data networks, will continue to struggle with the concepts of integration, unification, and standards. The quality of the software supporting traffic monitoring and control, its ability to provide rapid and accurate fault diagnosis, and all functions vital to VSAT networking is a determining factor to VSAT network success. It is very likely that there will be continuous future developments in the field.

6.6 CONCLUSION

Table 6.1 summarises the above discussion by categorising evolution trends according to their likelihood over a period of time. Short term is envisaged on a time scale of five years, long term represents a time scale of ten years.

APPENDICES

APPENDIX 1   TRAFFIC SOURCE MODELS

The purpose of traffic source modelling is to provide mathematical tools that can be used in simulations, and also give some insight on the physical behaviour of the device acting as a source. For any network design and performance evaluation, it is important to have models. Such models should be matched to the actual traffic, measured at these interfaces.

The model should include the following features:

—the rate at which messages are generated,

—the time between messages, also named interarrival time,

—the duration (in seconds) or length (in bits) of the message.

The Poisson model is very popular and useful. The defining assumptions are as follows [SCH77, p. 105]:

—the probability of one message being generated in a small time interval Δt ($\Delta t \to 0$) is proportional to that interval, therefore equal to $\lambda \Delta t$, where λ is a constant;

—Δt is considered small enough, so there cannot be more than one arrival in Δt, and the probability of no message being generated in Δt is equal to $(1 - \lambda \Delta t)$.

Message generation

Based on these assumptions, it can be shown that the probability of K messages being generated in an interval T, much larger than Δt, is given by:

$$P(K) = (\lambda T)^K \frac{e^{-\lambda T}}{K!} \tag{A1.1}$$

The average number of messages being generated in T seconds is then:

$$\langle K \rangle = \lambda T \tag{A1.2}$$

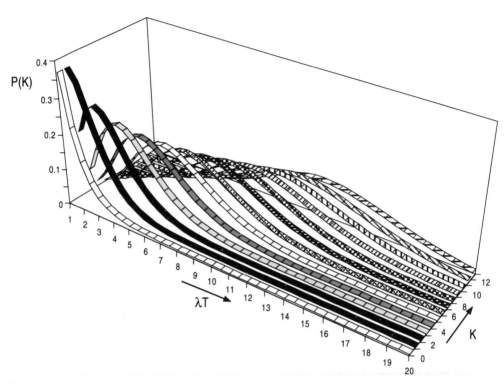

Figure A1.1   Probability $P(K)$ of K messages being generated during time T as a function of λT.

So, the Poisson parameter introduced as a proportionality factor for the small interval Δt turns out to be the average number of messages generated per unit time as well. Figure A1.1 represents the probability $P(K)$ of K messages being generated during time T as a function of λT. It reaches a maximum equal to $(K/e)^K/K!$ for $\lambda T = K$.

Interarrival time (IAT)

It can also be shown that the time τ between messages is a continuously distributed exponential random variable:

$$f(\tau) = \lambda e^{-\lambda \tau} \qquad (s^{-1}) \tag{A1.3}$$

where $f(\tau)$ is the probability density function of τ, and is represented in Figure A1.2.

The average interarrival time is:

$$\langle \text{IAT} \rangle = \int_0^\infty \tau f(\tau)\, d\tau = \frac{1}{\lambda} \qquad (s) \tag{A1.4}$$

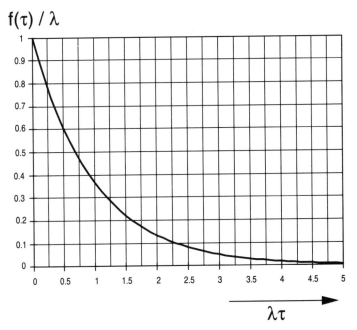

Figure A1.2   Normalised probability density function of message interarrival time.

Message length

Exponential distribution

The probability for a message to be L bits in length is given by:

$$P(L) = \mu e^{-\mu L} \qquad\qquad\qquad (A1.5)$$

The average message length is:

$$\langle L \rangle = \frac{1}{\mu} \qquad\qquad\qquad (A1.6)$$

Geometric distribution

The probability for a message to be L bits in length is given by:

$$P(L) = p q^{L-1} \qquad\qquad\qquad (A1.7)$$

where p is the probability of having a one-bit message, and $q = 1 - p$.
    The average message length is:

$$\langle L \rangle = \frac{1}{p} \qquad\qquad\qquad (A1.8)$$

APPENDIX 2 AUTOMATIC REPEAT REQUEST (ARQ) PROTOCOLS

Notation

D = number of data bits per frame to be conveyed from source to destination
L = length of frame (bits) = $D + H$
H = total number of bits in the frame header (and trailer if any)
R_b = information bit rate over the connection (b/s)
BER = bit error rate (s^{-1})
P_f = probability for a frame to be in error = $1 - (1 - \text{BER})^L$
T_{RT} = round trip time = $L/R_b + T_p + A/R_b + T_p = (A + L)/R_b + 2T_p \approx L/R_b + 2T_p$
T_P = propagation time from link source node to link destination node
A = length (in bits) of acknowledgement frame (considered negligible compared to L)

Channel efficiency

The channel efficiency is defined as the ratio of the throughput THRU to the information bit rate R_b of the source:

$$\eta_c = \frac{\text{THRU}}{R_b} \tag{A2.1}$$

Stop-and-wait (SW) protocol

The delivery time of a frame is equal to the average number $\langle M_{SW} \rangle$ of transmissions multiplied by the round trip time T_{RT} between each transmission (one error free frame is transmitted each T_{RT} interval). Therefore, the throughput is equal to:

$$\text{THRU} = \frac{D}{T_{RT} \langle M_{SW} \rangle} \quad \text{(b/s)} \tag{A2.2}$$

and the channel efficiency is given by

$$\eta_{cSW} = \frac{D}{R_b \, T_{RT} \langle M_{SW} \rangle} \tag{A2.3}$$

The probability $P(k)$ for a frame to be transmitted k times is having a frame correctly received after $k - 1$ unsuccessful tries:

$$P(k) = (1 - P_f)P_f^{k-1} \tag{A2.4}$$

The average number of transmission $\langle M_{SW} \rangle$ is equal to:

$$\langle M_{SW} \rangle = \sum_{k=0}^{\infty} kP(k)$$

$$= (1 - P_f) \sum_{k=0}^{\infty} kP_f^{k-1}$$

$$= (1 - P_f) \frac{d}{dP_f} \left(\sum_{k=0}^{\infty} P_f^k \right)$$

$$= (1 - P_f) \frac{d}{dP_f} \left(\frac{1}{1 - P_f} \right)$$

$$= \frac{1}{1 - P_f} \tag{A2.5}$$

Therefore:

$$\eta_{cSW} = \frac{D(1 - P_f)}{R_b T_{RT}} \tag{A2.6}$$

Go-back-N (GBN) protocol

Each NACK received by the sender initiates the retransmission of all subsequent frames transmitted after the non-acknowledged one. Let us assume that the sender, upon reception of an ACK, is allowed to transmit W frames and stop until reception of the next ACK. If a NACK is received, the number of frames that are retransmitted is W.

The delivery time of a frame is equal to the average number $\langle M_{GBN} \rangle$ of transmissions of that frame multiplied by its duration, L/R_b. Hence, the channel efficiency is given by:

$$\eta_{cGBN} = \frac{D}{L \langle M_{GBN} \rangle} \tag{A2.7}$$

Table A2.1

Number of transmitted frames	Number of unsucessful transmissions E	Probability
1	0	$P(E = 0) = (1 - P_f)$
$W + 1$	1	$P(E = 1) = (1 - P_f)P_f$
$2W + 1$	2	$P(E = 2) = (1 - P_f)P_f^2$
...	...	....
$kW + 1$	k	$P(E = k) = (1 - P_f)P_f^k$

The probabilities associated with the number of transmissions of a given frame are given in Table A2.1.

The average number of transmissions of a given frame is equal to:

$$\langle M_{GBN} \rangle = \sum_{k=0}^{\infty} (kW + 1)(1 - P_f)P_f^k$$

$$= W(1 - P_f) \sum_{k=0}^{\infty} kP_f^k + \sum_{k=0}^{\infty} (1 - P_f)P_f^k$$

Considering

$$\sum_{k=0}^{\infty} kP_f^k = P_f \sum_{k=0}^{\infty} kP_f^{k-1}$$

$$= P_f \frac{d}{dP_f} \left(\sum_{k=0}^{\infty} P_f^k \right)$$

$$= P_f \frac{d}{dP_f} \left(\frac{1}{1 - P_f} \right)$$

$$= \frac{P_f}{(1 - P_f)^2}$$

and

$$\sum_{k=0}^{\infty} (1 - P_f)P_f^k = \sum_{k=0}^{\infty} (P_f^k - P_f^{k+1}) = 1$$

then

$$\langle M_{GBN} \rangle = WP_f/(1 - P_f) + 1 = (1 + (W - 1)P_f)/(1 - P_f) \qquad (A2.8)$$

One can verify that $W = 1$ corresponds to the stop-and-wait case, as given by (A2.5).

According to (A2.7):

$$\eta_{cGBN} = \frac{D}{L} \frac{1 - P_f}{1 + (W - 1)P_f} \qquad (A2.9)$$

The maximum value of η_{cGBN} is obtained for continuous transmission. This occurs if the window W exceeds the maximum number of frames on the link. A frame lasts L/R_b and the link can hold $T_{RT}/(L/R_b)$ frames. Therefore, the condition for continuous transmission is:

$$W > \frac{R_b T_{RT}}{L} \qquad (A2.10)$$

The channel efficiency is bounded by:

$$\eta_{cGBN} < \frac{D}{L} \frac{1 - P_f}{1 + \left(\dfrac{R_b T_{RT}}{L} - 1 \right)P_f}$$

$$= \frac{D(1 - P_f)}{L(1 - P_f) + T_{RT} R_b P_f} \qquad (A2.11)$$

Selective-repeat (SR) protocol

Upon reception of a NACK, the sender retransmits the erroneous frame only. At the receiver, frames must be stored and re-sequenced.

The delivery time of a frame is equal to the average number $\langle M_{SR} \rangle$ of transmissions of that frame multiplied by its duration L/R_b. $\langle M_{SR} \rangle$ is equal to $\langle M_{SW} \rangle$. Hence:

$$\langle M_{SR} \rangle = \frac{1}{1 - P_f} \qquad (A2.12)$$

$$\eta_{cSR} = \frac{D(1 - P_f)}{L} \qquad (A2.13)$$

APPENDIX 3   INTERFACE PROTOCOLS

This appendix intends to give brief information on the most common protocols. For more details, the reader is invited to refer to the specialised literature, such as for instance [TUG82][TAN89] and the relevant texts of the EIA (Electronic Industries Association) and the ITU-T (formerly CCITT).

ASYNC (Asynchronous Communications)

Each information character or block of data is individually synchronised, usually by the use of start and stop elements. The gap between each character or block is not necessarily of a fixed length. Asynchronous data are usually produced by low speed terminals with bit rates up to a few kb/s.

BISYNC (Binary Synchronous Communications), also termed BSC

A set of control characters and control character sequences for synchronised transmission of binary-coded data between devices in a data communications system (see HDLC).

HDLC (High Data Level Protocol)

A Layer 2 (data link) protocol which rules orderly transfer of information between interfaced computers or terminals. The basic functions of HDLC are:

—to establish and terminate a connection between two terminals;

—to assure the message integrity through error detection, request for retransmission, and positive or negative acknowledgements;

—to identify the sender and the receiver through polling or selection;

—to handle special control functions such as requests for status, station reset, reset acknowledgement, start, start acknowledgement, and disconnection.

PAD (Packet Assembler/Disassembler)

Applies to exchange of serial data streams with character-mode terminal and the packetising–depacketising of the corresponding data exchanged with the ITU-T X25 terminal. Among the basic functions of the PAD are:

—assembly of characters into packets destined for the X25 Data Terminating Equipment (DTE);

—disassembly of the user data field of packets destined for the start–stop mode DTE (asynchronous transmission in which a group of code elements corresponding to a character signal is preceded by a start element and followed by a stop element);

—handling of virtual call set-up and clearing, resetting and interrupt procedures;

—generation of service signals;

—a mechanism for forwarding packets when the proper conditions exist, such as when a packet is full or an idler timer expires;

—a mechanism for transmitting data characters, including start, stop, and parity elements as appropriate to the start–stop DTE;

—a mechanism for handling a 'break' signal from the start–stop DTE.

RS232

A Layer 1 (physical layer) protocol standard as well as an electrical standard specifying handshaking functions between the Data Terminating Equipment (DTE) and the Data Circuit Terminating Equipment (DCE) over short distances (up to 15 m) at low-speed data rates (upper limit of 20 kb/s).

RS232 makes use of a 25-pin connector. A positive voltage between $+5$ and $+25$ V represents a logic 0, and a negative voltage between -5 and -25 V represents logic 1.

The ITU-T counterparts of RS232 are V24 and V28.

RS422

Layer 1 (physical layer) protocol standard.

It is a differential balanced voltage interface standard capable of higher data rates over longer distances than those specified in RS232.

The ITU-T counterparts of RS422 are Recommendations V11 and X27.

RS449

Expands specifications of RS232 to higher data rates and longer distances (for instance, 2 Mb/s over 60 m cables). The mechanical, functional and procedural interfaces are given in RS449, but the electrical interfaces are given by RS423 for unbalanced transmission (all circuits share a common ground) and RS422 for balanced transmission (each one of the circuits requires two wires with no common ground). RS449 makes use of a 37 pin connector.

SDLC (Synchronous Data Link Control)

An IBM variant of HDLC.

SNA/SDLC

See SDLC.

TCP/IP (Transmission Control Protocol/Internet Protocol)

A set of protocols from layer 3 to 5 (network/transport/session layers) developed to allow cooperating computers to share resources across a network. Among the basic functions of TCP/IP are:

—file transfer;

—remote login;

—computer mail;

—access to distributed databases, etc.

V11 (also X27)

Deals with the electrical characteristics of balanced double current interchange circuits operating with data signalling rates up to 10 Mb/s. It is similar to RS422.

V24

A list of definitions for interchange circuits between Data Terminating Equipment (DTE) and Data Circuit Terminating Equipment (DCE) for transfer of binary data, control and timing signals. The definitions are applicable to synchronous and asynchronous data communications. It is similar to RS232.

V28

Defines the electrical characteristics for unbalanced double-current interchange circuits operating below 20 kb/s. Binary 1 corresponds to voltages lower than -3 V. Binary 0 corresponds to voltages higher than $+3$ V. It is similar to RS232.

V35

Defines interchange circuits for data transmission at 48 kb/s. Practice has established V35 as a standard for interface circuits operating at 48, 56 and 64 kb/s using a 34-pin connector. It is similar to RS232, with slight differences (no external transmit clock).

X3

Describes the basic functions and the user-selectable functions of the Packet Assembler–Disassembler (PAD). It applies to exchange of serial data streams to/from a character-mode terminal from/to an X25 terminal (transport layer).

X21

Applies to the DTE/DCE interface for synchronous operation on public data networks. It defines functions at all three lower layers of the network.

X25

Defines a set of protocols for block transfer between a host computer and a packet switching network. For layer 1 (physical layer), X25 specifies layer 1 of X21. For layer 2 (data link layer), a link access protocol (LAP) is defined using the principles of high level data link control (HDLC). This layer provides the function of error and flow control for the access link between the DTE and the network. Each frame has a check sequence to detect errors, and error frames are retransmitted when requested by the receiving end or by time-out. Flow control is accomplished through the sending of receiver ready and receiver not ready commands. Layer 3 of X25 (network layer) defines the packet formats and control procedures for exchange of information between Data Terminating Equipment (DTE) and the network. X25 provides for the capability of multiplexing up to 4096 logical channels, or virtual circuits, on a single access link. Each channel can be used for virtual calls or a permanent virtual circuit. Each packet exchanged across the interface has its associated logical channel number identified, and each logical channel operates independently of the others. The data packets are also identified by sequence numbers which are used for flow control within individual logical channels. The sequence numbering may be based upon either modulo 8 for normal operation or modulo 128 for extended transmission delay conditions. Data packets are limited to a maximum data field length (nominally 128 octets, with possible extension up to 1024 octets).

X28

Defines the interface for the start–stop mode terminals accessing the Packet Assembler–Disassembler (PAD) on a public data network. It specifies procedures for establishing an access information path between a start–stop Data Terminating Equipment; (DTE) and a PAD, and for character interchange and service initialisation between them, as well as for the exchange of control information. It also summarises PAD commands and service signals.

X29

Specifies the procedures for the exchange of control information and user data between an X25 Data Terminating Equipment (DTE) and a Packet Assembler–Disassembler (PAD).

APPENDIX 4   ANTENNA PARAMETERS

Gain

Definition

The gain of an antenna is the ratio of the power radiated (or received) per unit solid angle by the antenna in a given direction to the power radiated (or received) per unit solid angle by an isotropic antenna fed with the same power.

Maximum gain

The gain is maximum in the direction of maximum radiation (the electromagnetic axis of the antenna, also called the *boresight*) and has a value given by:

$$G_{max} = \frac{4\pi}{\lambda^2} A_{eff} \tag{A4.1}$$

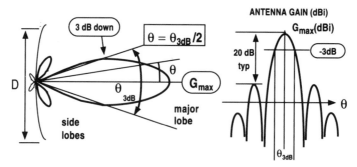

Figure A4.1 Antenna radiation pattern: (a) polar coordinates; (b) cartesian coordinates (dB on vertical scale).

where $\lambda = c/f$; c is the velocity of light ($c = 3 \times 10^8$ m/s) and f is the frequency of the electromagnetic wave. A_{eff} is the *effective aperture area* of the antenna. For an antenna with a circular aperture or reflector of diameter D and geometric surface $A = \pi D^2/4$, $A_{\text{eff}} = \eta_a A$, where η_a is the efficiency of the antenna (a typical value for antenna technology and frequencies used in VSAT networks is $\eta_a = 0.6$). Hence:

$$G_{\text{max}} = \eta_a \left(\frac{\pi D}{\lambda}\right)^2 = \eta_a \left(\frac{\pi Df}{c}\right)^2$$

$$G_{\text{max}}(\text{dBi}) = 10 \log\left[\eta_a \left(\frac{\pi D}{\lambda}\right)^2\right] = 10 \log\left[\eta_a \left(\frac{\pi Df}{c}\right)^2\right] \qquad \text{(A4.2)}$$

dBi means relative to isotropic antenna.

Antenna radiation pattern

The *antenna radiation pattern* indicates the variations of gain with direction. For an antenna with a circular aperture or reflector, this pattern has rotational symmetry about its boresight and can be represented by its variation within any plane containing the boresight. Figure A4.1 displays a typical pattern which can be represented either in polar coordinates (Figure A4.1a) or in cartesian coordinates (Figure A4.1b).

Figure A4.1 reveals the major lobe which contains the direction of maximum gain G_{max} at boresight ($\theta = 0°$) and the side lobes with smaller secondary maxima.

Half power beamwidth

It is convenient to characterise the width of the antenna radiation pattern by the angle between the direction in which the gain falls to half its maximum value. This angle is called the *3dB beamwidth* θ_{3dB}. A practical formula for θ_{3dB} is:

$$\theta_{3dB} = 70\frac{\lambda}{D} = 70\frac{c}{fD} \qquad \text{(degrees)} \qquad (A4.3)$$

It can be noted that θ_{3dB} increases with decreasing D, which indicates that a small aperture antenna displays a large beamwidth.

Depointing loss

In a direction θ near to the boresight, say between 0 and $\theta_{3dB}/2$, the value of the gain is given by:

$$G(\theta)(\text{dBi}) = G_{max}(\text{dBi}) - 12\left(\frac{\theta}{\theta_{3dB}}\right)^2 \qquad \text{(dBi)} \qquad (A4.4)$$

Polarisation

The wave radiated by an antenna consists of an electric field component and a magnetic field component. These two components are orthogonal and perpendicular to the direction of propagation of the wave. They vary in time at the frequency of the wave. By convention the *polarisation* of the wave is defined by the direction of the electric field. In general, the direction of the electric field is not fixed and its amplitude is not constant. During one period, the projection of the extremity of the vector representing the electric field onto a plane perpendicular to the direction of propagation of the wave describes an ellipse, as illustrated in Figure A4.2. The polarisation is said to be elliptical.

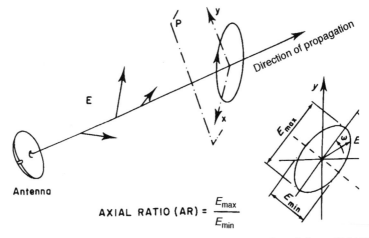

$$\text{AXIAL RATIO (AR)} = \frac{E_{max}}{E_{min}}$$

Figure A4.2   Polarisation of an electromagnetic wave. (Reproduced from [MAR93] by permission of John Wiley & Sons Ltd.)

Two waves are in *orthogonal polarisation* if their electrical fields describe identical ellipses in opposite directions. In particular, the following can be obtained:

—two orthogonal circular polarisations are described as right-hand circular and left-hand circular (the direction of rotation is for an observer looking in the direction of propagation);

—two orthogonal linear polarisations are described as horizontal and vertical (relative to a local reference).

An antenna designed to transmit or receive a wave of given polarisation cannot transmit or receive in the orthogonal polarisation. This allows two simultaneous links to be set up at the same frequency between the same two locations, so called 'frequency reuse' by orthogonal polarisation. To achieve this, either two polarised antennas are installed at each end or, preferably, one antenna designed for operation with the two specified polarisations may be used. This practice must, however, take account of imperfections of the antennas and the possible depolarisation of the waves by the transmission medium (the atmosphere, especially with rain, in the case of satellite links). These effects introduce mutual interference of the two links. This situation is illustrated in Figure A4.3 which relates to the case of two orthogonal linear polarisations (the illustration is also valid for any two orthogonal polarisations).

Let a and b be the amplitudes, assumed to be equal, of the electric field of the two waves transmitted simultaneously with linear polarisation, a_c and b_c be the amplitudes received with the same polarisation, and a_x and b_x be the amplitudes received with orthogonal polarisations. The following concepts are defined:

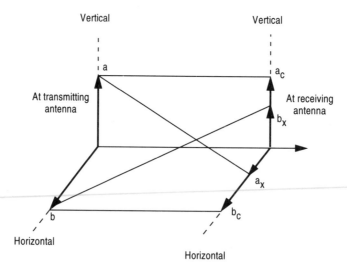

Figure A4.3 Amplitude of the transmitted and received electric field for the case of two orthogonal polarisations.

—the *cross-polarisation isolation*:

$$\text{XPI} = \frac{a_c}{b_x} \quad \text{or} \quad \frac{b_c}{a_x}$$

$$\text{XPI(dB)} = 10\log\left(\frac{a_c}{b_x}\right) \quad \text{or} \quad 10\log\left(\frac{b_c}{a_x}\right) \qquad \text{(A4.5)}$$

—the *cross-polarisation discrimination* (when a single polarisation is transmitted):

$$\text{XPD} = \frac{a_c}{a_x} \qquad \text{(A4.6)}$$

In practice, XPI and XPD are comparable and are often confused within the term *isolation*.

The values and relative values of the components vary as a function of the direction relative to the antenna boresight. The antenna is thus characterised, for a given polarisation, by a radiation pattern for the nominal polarisation (co-polar) and a radiation pattern for the orthogonal polarisation (cross-polar). Cross-polarisation discrimination is usually maximum on the antenna axis and degrades for directions other than boresight.

APPENDIX 5   EMITTED AND RECEIVED POWER

Effective isotropic radiated power (EIRP) of a transmitter

The power radiated per unit solid angle by an isotropic antenna fed from a radio frequency source of power P_T is given by $P_T/4\pi$ (W/steradian).

In a direction where the value of transmission gain is G_T, any antenna radiates a power per unit solid angle equal to $G_T(P_T/4\pi) = (P_TG_T)/4\pi$ (W/steradian).

The product P_TG_T is called the *effective isotropic radiated power* (EIRP). It is expressed in W.

$$\begin{aligned} \text{EIRP} &= P_TG_T \qquad\qquad\qquad \text{(W)} \\ \text{EIRP(dBW)} &= P_T\text{(dBW)} + G_T\text{(dBi)} \end{aligned} \qquad \text{(A5.1)}$$

Power flux density at receiver

A surface of effective area A situated at a distance R from the transmitting antenna subtends a solid angle A/R^2 at the transmitting antenna. It receives a power equal to the product of the power radiated per unit solid angle $(P_TG_T)/4\pi$ and the

ISOTROPIC ANTENNA

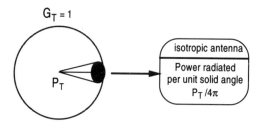

ACTUAL ANTENNA

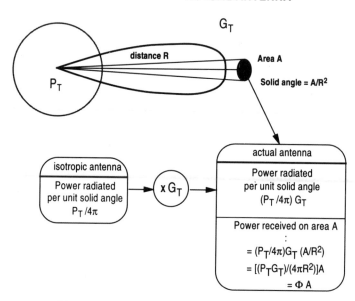

Figure A5.1   Power flux density at receiver.

considered solid angle A/R^2:

$$P_R = \frac{P_T G_T}{4\pi}\frac{A}{R^2} = \Phi A \qquad (W)$$

$$P_R(\text{dBW}) = \Phi(\text{dBW}/\text{m}^2) - 10\log A$$

$(A5.2)$

where Φ is called the *power flux density*:

$$\Phi = \frac{P_T G_T}{4\pi R^2} \qquad (W/m^2)$$

$$\Phi(\text{dBW}/\text{m}^2) = \text{EIRP}(\text{dBW}) - 10\log(4\pi R^2)$$

$(A5.3)$

Figure A5.1 summarises the above derivations.

Power available at the output of the receiving antenna

Figure A5.2 represents a receiving antenna of effective area A_{Reff} located at a distance R from a transmitting antenna.

From (A5.2), the receiving antenna captures a power equal to:

$$P_R = \Phi A_{Reff} = \frac{P_T G_T}{4\pi} \frac{A_{Reff}}{R^2} \quad (W) \tag{A5.4}$$

The equivalent area of an antenna is expressed as a function of its receiving gain G_R by the expression:

$$A_{Reff} = \frac{G_R}{4\pi/\lambda^2} \quad (m^2) \tag{A5.5}$$

The ratio $4\pi/\lambda^2$ can be looked upon as the gain of an ideal antenna with an area equal to $1\,m^2$.

Denoting $G_1 = 4\pi/\lambda^2$:

$$A_{Reff} = \frac{G_R}{G_1} \quad (m^2) \tag{A5.6}$$

From (A5.4) and (A5.6), one can derive an expression for the received power:

$$P_R = \Phi A_{Reff} = \Phi\left(\frac{G_R}{G_1}\right) \quad (W) \tag{A5.7}$$

$$P_R(dBW) = \Phi(dBW/m^2) + G_R(dBi) - G_1(dBi) \quad (dBW)$$

From (A5.4) and (A5.6), one can derive another expression for the received power:

$$P_R = \frac{P_T G_T}{L_{FS}} G_R \quad (W) \tag{A5.8}$$

$$P_R(dBW) = EIRP(dBW) - L_{FS}(dB) + G_R(dBi) \quad (dBW)$$

where $L_{FS} = (4\pi R/\lambda)^2$ is called the *free space loss*, and represents the ratio of the transmitted to the received power in a link between two isotropic antennas.

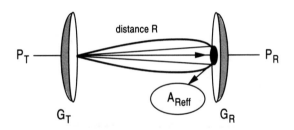

Figure A5.2   Power captured by a receiving antenna.

APPENDIX 6   CARRIER AMPLIFICATION

Carrier amplification takes place at the transmitting earth station (a VSAT or the hub) and on board the satellite, within each transponder. Power amplifiers used are either solid state power amplifiers (SSPA) or Travelling Wave Tubes (TWT). Both types act as non-linear devices when operated near saturation, where the ouput power is maximum. The non-linearity has two aspects: a decreasing power gain, as the output power comes to saturation, and a variation in the phase of the amplified carrier relative to the input phase.

Figure A6.1 displays typical transfer characteristics for a power amplifier. All quantities are normalised to their respective values at saturation, when the amplifier is operated in a single carrier mode.

Denoting by P_0^1 the output power, and by P_i^1 the input power (1 stands for single carrier drive, 0 for output, i for input), $(P_0^1)_{sat}$ and $(P_i^1)_{sat}$ those quantities at saturation, one defines the output back-off (OBO) and the input back-off (IBO) as:

$$\text{OBO} = \frac{P_0^1}{(P_0^1)_{sat}}$$

(A6.1)

$$\text{OBO(dB)} = 10\log\left\{\frac{P_0^1}{(P_0^1)_{sat}}\right\}$$

$$\text{IBO} = \frac{P_i^1}{(P_i^1)_{sat}}$$

(A6.2)

$$\text{IBO(dB)} = 10\log\left\{\frac{P_i^1}{(P_i^1)_{sat}}\right\}$$

The values available from the amplifier manufacturer are the output power at saturation $(P_0^1)_{sat}$ and the power gain at saturation G_{sat}. From those two quantities, one can derive $(P_i^1)_{sat}$ as:

$$(P_i^1)_{sat} = \frac{(P_0^1)_{sat}}{G_{sat}}$$

(A6.3)

For example, a 50 W TWT, with a 55 dB gain, displays an input power at saturation $(P_i^1)_{sat} = 50\,\text{W}/10^{5.5} = 158\,\mu\text{W}$.

With the above definitions, the values for OBO(dB) and IBO(dB) are negative in the normal range of operation, i.e. below saturation. Note that some people define OBO and IBO as the inverse of expressions (5.53) and (5.54). OBO (dB) and IBO (dB) values are then positive.

The aspect of the curves in Figure A6.1 is non-linear. When operated in a multicarrier mode, the non-linearity generates intermodulation, and the amplifier output power is shared not only between the amplified carriers but also with the intermodulation products (see section 5.1.3). Denoting by P_0^n and P_i^n, respectively, the output and input power of one carrier among the n amplified ones, one can define:

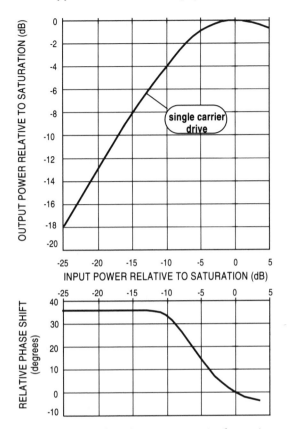

Figure A6.1 Power amplifier characteristic: single carrier operation.

—the output back-off *per carrier*:

OBO$_1$ = single carrier output power/single carrier output power at saturation

$$= \frac{P_0^n}{(P_0^1)_{sat}} \tag{A6.4}$$

$$\text{OBO}_1(dB) = 10 \log \left\{ \frac{P_0^n}{(P_0^1)_{sat}} \right\}$$

—the input back-off *per carrier*:

IBO$_1$ = single carrier input power/single carrier input power at saturation

$$= \frac{P_i^n}{(P_i^1)_{sat}} \tag{A6.5}$$

$$\text{IBO}_1(dB) = 10 \log \left\{ \frac{P_i^n}{(P_i^1)_{sat}} \right\}$$

—the *total* output back-off:

OBO_1 = sum of all carrier output power/single carrier input power at saturation

$$= \frac{\Sigma P_0^n}{(P_0^1)_{sat}} \tag{A6.6}$$

$$OBO_t(dB) = 10 \log \left\{ \frac{\Sigma P_0^n}{(P_0^1)_{sat}} \right\}$$

—the *total* input back-off:

IBO_t = sum of all carrier input power/single carrier input power at saturation

$$= \frac{\Sigma P_i^n}{(P_i^1)_{sat}} \tag{A6.7}$$

$$IBO_t(dB) = 10 \log \left\{ \frac{\Sigma P_i^n}{(P_i^1)_{sat}} \right\}$$

With n equally powered carriers:

$$OBO_1 = \frac{OBO_t}{n}$$

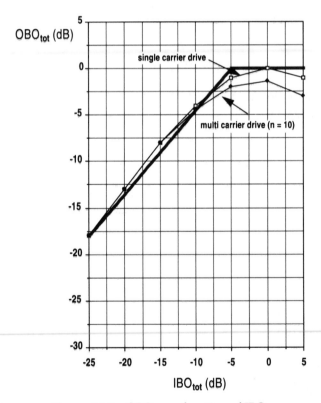

Figure A6.2 OBO_t as a function of IBO_t.

$$OBO_1(dB) = OBO_t(dB) - 10 \log n \qquad (A6.8)$$

$$OBO_1 = \frac{OBO_t}{n}$$

$$IBO_1(dB) = IBO_t(dB) - 10 \log n \qquad (A6.9)$$

Figure A6.2 gives typical variations of OBO_t as a function of IBO_t. A simple but useful model consists in approximating the curves by the two shaded segments:

$$
\begin{array}{ll}
OBO_t(dB) = 0.9(IBO_t(dB) + 5) & IBO_t < -5\,dB \\
OBO_t(dB) = 0\,dB & -5\,dB < IBO_t < 0\,dB
\end{array}
\qquad (A6.10)
$$

For the EUTELSAT 2 satellite, the following formula can be used [EUT92]:

$$OBO_t(dB) = -1.7 - 0.0313(IBO_t(dB))^2 \qquad -13\,dB < IBO_t < 0\,dB \quad (A6.11)$$

APPENDIX 7   VSAT PRODUCTS

This appendix aims at a survey of some of the existing VSAT product. The list of presented products is by no mean exhaustive, and information is subject to change. It is therefore recommended that the reader refers to the most recent information released by the manufacturer.

Hughes Network Systems (HNS)

Address: 11717 Exploration Lane, Germantown, MD 20878, USA
Phone: +1 301 428 5500
Fax: +1 301 428 1868

Equipment name	PES 6000	PES800
Application	Interactive data, for star shaped private satellite VSAT networks	
Satellite interface		
Operating band	Ku-band	Ku-band
Size of antenna (m)	0.75–1.0 m	1–1.2–1.8–2.4 m
Transmitter power (W)	0.5–1 W	0.5–1–2 W
Inbound transmit information rate	64 kb/s	64–128 kb/s
Outbound receive information rate	128 kb/s	128–256–512 kb/s
Access techniques	S-ALOHA/DA/STREAM	S-ALOHA/DA/STREAM

User interface		
Number of ports	2 to 8	4 to 32
Type of ports	RS232–RS422–V35	RS232–RS422–V35 & RJ11 (voice)
Data rate	0–19.2 kb/s async	0–19.2 kb/s async
	1.2–64 kb/s sync	1.2–64 kb/s sync
Protocols	SDLC	SDLC
	X25	X25
	Burrough's Poll Select	Burrough's Poll Select
	BSC 3270	BSC 3270
	X3/X28 PAD	X3/X28 PAD
Bit error rate to user	10^{-7}	10^{-7}

Operational conditions		
Operating temperature	IDU: 10° to + 40°C	IDU: 10° to + 40°C
	ODU: − 30° to + 55°C	ODU: − 30° to + 55°C

Equipment name	Telephony earth station
Application	Voice and data on thin route public or private networks with meshed configuration

Satellite interface		
Operating band	C-band	Ku-band
Access techniques	SCPC/DAMA/FDMA	SCPC/DAMA/FDMA

User interface		
Number of ports	up to 4 full duplex channels per chassis, up to six chassis per rack	up to 4 full duplex channels per chassis, up to six chassis per rack
Data rate	up to 19.2 kb/s async	0–19.2 kb/s async
	up to 56 kb/s sync	1.2–64 kb/s sync

Operational conditions		
Operating temperature	IDU: 0° to + 40°C	IDU: 10° to + 40°C

Equipment name	Gemini		
Application	Data, voice and video in point to point or star network configuration		
Satellite interface			
Operating band		Ku-band	
Size of antenna (m)	1.8	2.4	3.5
Receive G/T (dBK^{-1})	21 dBK^{-1}	23.3 dBK^{-1}	27.1 dBK^{-1}
Transmit EIRP (dBW)	46 dBW	48 dBW	51.6 dBW
Polarisation		Orthogonal linear	
Transmitter power (W)		1.5 or 3W	
Link information rate (kb/s)	up to 56 kb/s	56–192 kb/s	192–256 kb/s
FEC code		1/2	
Channel spacing		200 kHz	
Access techniques		BPSK (56 kb/s) QPSK > (56 kb/s)	
User interface			
Type of ports	RS499/422 standard (V35 optional)		
Data rates	Single or dual 56 kb/s, Single 112/kb/s, Single 384 kb/s, Single 512 kb/s		
Bit error rate to user	10^{-7}		
Operational conditions			
Operating windspeed		up to 80 km/h	
Survival windspeed		up to 200 km/h	

GTE Spacenet

Address: 1750 Old Meadow Rd, McLean, VA 22102, USA
Phone: +1 703 848 1000
Fax: +1 703 848 0004

GTE Spacenet has worked with 12 VSAT vendors since 1984 and currently operates networks involving different types of VSATs. GTE Spacenet has predominantly used NEC's Ku-band VSAT equipment with a proprietary software overlay. The most recent network offering of GTE is named SKYSTAR Plus. This is a star shaped network intended for use in multiple application, multiple protocol, interactive data networks which must handle both batch and interactive traffic. Another type of star network offered by GTE is the EQUASTAR, which is a low cost C-band VSAT system intended for interactive data, voice and facsimile applications. GTE has also been offering network services based on GILAT equipment.

ATT Tridom

Address: 840 Franklin Ct, Marietta, GA 30067, USA
Phone: +1 404 426 4201
Fax: +1 404 514 1737

Equipment	Clearlink Plus		Clearlink 200	Clearlink 400	
Application	Interactive data and voice		Interactive data	Interactive data and voice	
Satellite interface					
Operating band	C-band	Ku-band	Ku-band	C-band	Ku-band
Transmit frequency band	5.925–6.425 GHz	14–14.5 GHz	14–14.5 GHz	5.850–6.425 GHz	14–14.5 GHz
Transmit step size	100 kHz	100/125 kHz	100 kHz	100 kHz or 125 kHz	
Receive frequency band	3.7–4.2 GHz	11.7–12.2 GHz and 12.5–12.75 GHz	11.7–12.2 GHz	3.625–4.2 GHz	10.95–12.75 GHz
Receive step size	500 kHz	500 kHz	250 kHz	250 kHz	
Size of antenna	1.8m, 2.4m		0.95m, 1.2m	1.8m, 2.4m, 3.5 m	1.2m, 1.8m, 2.4m
Transmitter power (W)	2 W	0.5–1 W	0.5–1 W	2–5 W	0.5–2 W
Modulation	BPSK		BPSK	BPSK	
Transmit information bit rate	64 kb/s	32–64 kb/s	64 kb/s	32–64 kb/s	32–64 kb/s
Inbound FEC code	rate 2/3		rate 2/3	rate 2/3	
Receive information bit rate (kb/s)	512 kb/s		512 kb/s	128–256–512 kb/s	
Outbound FEC code	rate 2/3		rate 1/2	rate 1/2	
Access techniques	ALOHA C/TDMA* FA/TDMA** CA/TDMA*** DC/TDMA****		ALOHA C/TDMA* FA/TDMA** CA/TDMA*** DC/TDMA****	ALOHA C/TDMA* FA/TDMA** CA/TDMA*** DC/TDMA****	

User interface

Number of ports	3	2	2 or 5
Type of ports	RS232–RS422–V35	RS232–RS530–V35	RS232–RS530–V35
Data rates		300 b/s–19.2 kb/s ASYNC	
		300 b/s–19.2 kb/s SYNC	
Protocols	3270 SNA/SDLC	3270 SNA/SDLC	3270 SNA/SDLC
	X25	X25	X25
	TCP/IP	TCP/IP	TCP/IP
	ASYNC	ASYNC	ASYNC
	3270 BISYNC	3270 BISYNC	3270 BISYNC
	NCR 751 BISYNC	NCR 751 BISYNC	NCR 751 BISYNC
	digital vocoder		digital vocoder

Operational conditions

Operating	IDU: 0° to +40°C
Temperature	ODU: −40° to +60°C

TDM: Time Division Multiplex, ALOHA: random access, used for light interactive traffic
*C/TDMA: Contention Time Division Multiple Access, used for light interactive traffic
**FA/TDMA: Fixed Assigned Time Division Multiple Access, used for heavy batch or dedicated bandwith
***CA/TDMA: Combined Access Time Division Multiple Access, used for light interactive and dedicated bandwidth.
****DC/TDMA: Demand Controlled TDMA, dynamically adjusts for simultaneous interactive and batch

Scientific Atlanta

Address: 420 N. Wickham Rd, Melbourne, FL 32935, USA
Phone: +1 407 255 3000
Fax: +1 407 259 3942

Equipment name	Skylin X.25
Applications	Interactive and batch data
	Voice, audio and data broadcast.

Satellite interface

Operating band	Ku-band	C-band
Transmit frequency band	14.0–14.5 GHz	5.925–6.425 GHz
Transmit step size	100 kHz	100 kHz
Receive frequency band	10.95–11.20 GHz	3.7–4.2 GHz
	11.7–12.2 GHz	
	12.25–12.75 GHz	
Receive step size	100 kHz	100 kHz
Size of antenna (m)	1.8 m	1.8 m
Transmitter power (W)	1 W	5 W
Modulation	BPSK	BPSK
Transmit information rate (kb/s)	112 or 128 kb/s	112 or 128 kb/s
Inbound FEC code	rate 1/2	rate 1/2
Transmit information receive rate (kb/s)	112 or 128 kb/s	112 or 128 kb/s
Outbound FEC code	rate 1/2	rate 1/2
Access techniques	S-ALOHA*	S-ALOHA*
	TDMA**	TDMA**
	Adaptive Reservation***	Adaptive Reservation***

User interface

Number of ports	4 or 8	4 or 8
Type of ports	RS232–RS422/423–RS449 or V35	RS232–RS422/423–RS449 or V35
Port speed	up to 19.2 kb/s	up to 19.2 kb/s
Data rates	1.2, 2.4, 4.8, 7.2, 9.6 or 19.2 kb/s	1.2, 2.4, 4.8, 7.2, 9.6 or 19.2 kb/s
Protocols	SNA/SDLC	SNA/SDLC
	BSC327X, 2780, 3780	BSC327X, 2780, 3780
	X25 levels 2 and 3	X25 levels 2 and 3
	X3, X28, X29 async	X3, X28, X29 async
	polled async	polled async
Bit error rate to user	10^{-7}	10^{-7}

Operational conditions

Operating temperature	IDU: 0° to 45°C	IDU: 0° to 45°C
	ODU: $-40°$ to $+55°$C	ODU: $-40°$ to $+55°$C

*Slotted ALOHA for light interactive traffic
**for heavy batch traffic or dedicated bandwidth
***dynamically adjusts for simultaneous interactive and batch traffic

Satellite Technology Management (STM)

Address: 3530 Hyland Ave., Costa Mesa, CA 92626, USA
Phone: +1 714 557 2400
Fax: +1 714 557 4239

Equipment name	X.STAR
Applications	Interactive and batch data

Satellite interface

Operating band	Ku-band
Size of antenna (m)	1.8 m
Transmitter power (W)	1–2 W
EIRP (dBW)	46.5 dBW
G/T (dBK^{-1})	23.5 dBK^{-1}
Modulation	BPSK
Transmit information rate	48–96–192 kb/s
Inbound FEC code	rate 1/2
Receive information rate	32 kb/s–7.5 Mb/s
Outbound FEC code	rate 1/2
Access techniques	TDMA
Bit error rate on satellite link	10^{-7}

User interface

Number of ports	5 to 17
Type of ports	RS232C (V35 opt)
Port speed	up to 64 kb/s
Protocols	SDLC 3270
	X25
	HDLC
	X3/X28 PAD

Operational conditions

Weight	150 kg
Operating wind speed	up to 105 km/h
Survival wind speed	up to 200 km/h

NEC

Address: NEC America Inc., 14040 Park Center Rd, Herndon, VA 22071, USA
Phone: +1 703 834 4150
Fax: +1 703 481 6904

Equipment name	NEXSTAR-IV advanced AA/TDMA	
Applications	Two-way interactive and batch transmission for star network	

Satellite interface		
Operating band	Ku-band	C-band
Transmit frequency band	14.0–14.5 GHz	5.925–6.425 GHz
		or 5.85–6.35 GHz
Transmit step size	100 kHz	100 kHz
Receive frequency band	10.95–11.20 GHz	3.7–4.2 GHz
	or 11.45–11.70 GHz	or 3.625–4.125 GHz
	or 11.70–12.20 GHz	
	or 12.5–12.75 GHz	
Receive step size	100 kHz	
Size of antenna (m)	0.95–1.2–1.8–2.4 m	
Transmitter power (W)	0.5–1–2 W	3.3 W
Modulation	BPSK/QPSK	
Transmit information bit rate	64 kb/s	
Inbound FEC code	rate 1/2	
Receive information bit rate	64/128 kb/s TDM from hub	
Outbound FEC code	rate 1/2	
Access techniques	AA/TDMA	
Bit error rate on satellite link	10^{-7}	

User interface	
Number of ports	max16
Type of ports	RS232(up to 19.2 kb/s)
	V35-EIA530-LAN802.3
Data rates	50b/s to 64 kb/s
Protocols	X25
	SNA/SDLC
	TCP/IP

Operational conditions	
Operating temperature	IDU: 0° to + 35°C
	ODU: − 30° to + 55°C

Equipment name	NEXTAR-CLVR (Clear channel Variable Rate)
Applications	Two-way SCPC for bulk data transfer (bit transparent)

Satellite interface

	Ku-band	C-band
Operating band		
Transmit frequency band	14.0–14.5 GHz	5.925–6.425 GHz
		or 5.85–6.325 GHz
Transmit step size	25 kHz	25 kHz
Receive frequency band	11.70–12.20 GHz	3.7–4.2 GHz
	or 12.25–12.75 GHz	or 3.625–4.1 GHz
Receive step size	25 kHz	
Size of antenna (m)	1.8 m–2.4 m–3.6 m–4.5 m	
Transmitter power (W)	1.5 W–2 W–6 W	5 W
Modulation	QPSK for 56 to 2048 kb/s	
Transmit information bit rate (kb/s)	9.6 to 2048 kb/s	
Inbound FEC code	rate 1/2 for 9.6 to 2048 kb/s	
	rate 3/4 for 64 to 2048 kb/s	
	rate 7/8 for 192 to 2048 kb/s	
Receive information bit rate (kb/s)	same as transmission (no hub)	
Outbound FEC code	same as transmission (no hub)	
Access techniques	variable rate SCPC/FDMA	
Bit error rate on satellite link	10^{-7}	

User interface	
Number of ports	1 up to 2048 kb/s
Type of ports	RS422/449–V35–RS232
Port speed	max 56 kb/s
Data rates	9.6 to 2048 kb/s

Operational conditions	
Operating temperature	IDU: 0° to + 35°C
	ODU: − 30° to + 55°C

Equipment name	NEXTAR-VO (Digital Voice)	

Applications	Telephony	

Satellite interface		

Operating band	Ku-band	C-band
Transmit frequency band	14.0–14.5 GHz	5.925–6.425 GHz
		or 5.85–6.325 GHz
Receive frequency band	10.95—11.2 GHz	3.7–4.2 GHz
	or 11.45–11.70 GHz	or 3.625–4.1 GHz
	or 11.70–12.20 GHz	
	or 12.25–12.75 GHz	
Size of antenna (m)	1.2 m–1.8 m–2.4 m	
Transmitter power (W)	0.5 W–1 W–2 W	3.3 W–20 W
Modulation	QPSK	
Transmit information bit rate	35 kb/s	
Inbound FEC code	rate 1/2 or 3/4	
Receive information bit rate (kb/s)	same as transmit	
Outbound FEC code	same as transmit	
Access techniques	DAMA/SCPC	
Bit error rate on satellite link	10^{-6}	

User interface		

Type of ports	2 wire loop, 4-wire E&M with Bell type I-V
Port speed	32 kb/s ADPCM voice coding

Operational conditions	

Operating temperature	IDU: 0° to + 35°C
	ODU: −30° to + 55°C

GILAT

Address: 24a Habarzel St., Tel Aviv 69710, Israel
Phone: +972 3 499068
Fax: +972 3 6487429

Equipment name	Two WAY	
Applications	Interactive data	

Satellite interface		
Operating band	Ku-band	C-band
Transmit frequency band	14–14.5 GHz	5.85–6.425 GHz
Receive frequency band	10.95–11.7 GHz	3.7–4.2 GHz
	or 11.7–12.2 GHz	
	or 12.25–12.75 GHz	
Size of antenna (m)	0.9–1.2 m	1.8–2.4 m
Modulation	BPSK	
Transmit information rate (kb/s)	4.8–9.6–19.2 kb/s	
Inbound FEC code	rate 1/2 or 3/4	
Receive information rate (kb/s)	64 kb/s	
Outbound FEC code	rate 1/2 or 1/4	
Access techniques	Adaptive TDMA	
Bit error rate on satellite link	10^{-7}	

User Interface	
Number of ports	2 or 4
Type of ports	RS232, LAN interface
Port speed	50 b/s to 128 kb/s
Data rates	up to 19.2 kb/s
Protocols	X25
	SNA/SDLC
	Asynchronous X3

Operational conditions	
Operating temperature	IDU: 5° to +45°C
	ODU: −40° to +60°C
Operating wind speed	90 km/h
Survival wind speed	190 km/h

Equipment name	FaraWAY

Applications	Rural telephony with mesh network

Satellite Interface

Operating band	C and Ku-band
Size of antenna (m)	depending on satellite and frequency band
Modulation	QPSK
Information rate	13kb/s
Forward error correction	rate 3/4
Carrier control	Voice activation
Channel spacing	22.5 kHz
Access technique	DAMA/SCPC
Bit error rate	10^{-6}

User Interface

Number of ports	4
Data rates	13kb/s GSM voice encoding

Operational Conditions

Temperature	0°C to 50°C

Equipment name	One WAY

Applications	Data broadcast

Satellite Interface

Operating band	Ku-band	C-band
Receive frequency band	10.95–11.7 GHz	3.7–4.2 GHz
	11.7–12.2 GHz	
	12.25–12.75 GHz	
Size of antenna (m)	from 0.6 m and up	
Modulation	PSK	
Information rate	9.6–64 kb/s	
Forward error correction	rate 1/2 or 1/4	
Bit error rate	10^{-7}	

User Interface

Number of ports	up to 4
Type of ports	RS232
Data rates	9.6–64 kb/s

Operational Conditions	
Temperature	5°C to 45°C

Dornier

Address: Communications Systems, P.O. Box 1420, D-7990 Friedrichshafen, Germany
Phone: +49 7545 8 83 40
Fax: +49 7545 8 57 09

Under the designation Advanced Business Communications via Satellites (ABCS), Dornier has developed a VSAT product for interconnection of Local Area Networks (LAN), with a medium data rate of typically 2Mb/s.

Equipment name	ABCS
Applications	Interconnection of LANs with mesh network configuration

Satellite Interface	
Operating band	Ku-band
Transmit frequency band	14–14.5 GHz
Receive frequency band	12.5–12.75 GHz
Size of antenna (m)	1.8–2.4 m
Transmitter power (W)	5–8–16 W
Modulation	BPSK or QPSK
Information rate	512–1024–2048–4096 kb/s
Forward error correction	rate 1/2 or uncoded
Multiple access technique	TDMA with dynamic capacity assignment (fixed assignment of time slots for base load and reservation on demand of additional time slots with short reaction time)
Bit error rate	10^{-7}

User Interface	
Protocols	LAN Ethernet/IEEE 802.3

Operational Conditions	
Temperature	5°C to 45°C

Multipoint Communications

Address: Satellite House, Eastways Industrial Park, Witham, Essex CM8 3YQ, UK
Phone: +44 376 510881
Fax: +44 376 502233

Equipment name	SE 7506/7706		SE 7510/7710	
Applications	Voice and data		Point to point communications in mesh or star networks	
Satellite Interface				
Operating band	C-band	Ku-band	C-band	Ku-band
Transmit frequency band	5.85–6.425 GHz	14–14.5 GHz	5.85–6.425 GHz	14.0–14.5 GHz
Receive frequency band	3.65–4.2 GHz	10.95–11.7 GHz 12.5–12.75 GHz	3.625–4.20 GHz	10.95–11.7 GHz 12.5–12.75 GHz
Size of antenna (m)	1.5 m	1.5 m	1.8–2.4 m	1.8–2.4 m
Transmitter power (W)	1–20 W	1–20 W	1–5–10–20 W	1–5–10–20 W
EIRP (dBW)	45–55 dBW	45–55 dBW	38.2–45.2–48.2–51.2 dBW	45–52–55–58 dBW
LNB noise temperature	45 K	100 K	100 K	100 K
G/T (dBK^{-1})	14.5 dB/K	23.2 dB/K	15.5–18 dB/K	23.2–25.8 dB/K
Modulation	BPSK		QPSK/BPSK	
Transmit information rate (kb/s)	9.6–64 kb/s		64–2048 kb/s	
Inbound FEC code	rate 1/2		rate 1/2 or 3/4	
Receive information rate	208.4 kb/s		64–2048 kb/s	
Outbound FEC code	rate 1/2		rate 1/2 or 3/4	
Access techniques	DAMA/SCPC/FDMA DAMA/TDM from hub		SCPC/MCPC/MESH	
Bit error rate on satellite link	10^{-7}		$10^{-4}/10^{-3}$	
User Interface				
Number of ports	1		programmable multiplexer	
Type of ports			RS232–RS449/422-V35-G703	
Data rate	9.6 kb/s synchronous		9.6–64 kb/s	
Operational conditions				
Operating temperature	IDU:0°C to 40°C ODU: −30° to +50°C		IDU:0°C to 40°C ODU: −20° to +50°C	
Wind loading: operational survival	120 km/h 200 km/h		120 km/h 200 km/h	
Typical cost per VSAT	£15 000		15 000–25 000	

APPENDIX 8   SATELLITE NEWS GATHERING

This appendix aims at a survey of some of the existing VSAT product for satellite news gathering. The list of presented products is by no means exhaustive, and information is subject to change. It is therefore recommended that the reader refers to the most recent information released by the manufacturer.

Dornier

Address: Communications Systems, P.O. Box 1420, D-7990 Friedrichshafen, Germany
Phone: +49 7545 8 83 40
Fax: +49 7545 8 57 09

Type of station	Fly away	Fly away	Van mounted
Equipment name	Fly Away	Fly Away	Fly Away
Operating band	Ku-band	Ku-band	Ku-band
Transmit frequency band	14–14.5 GHz	14–14.5 GHz	14–14.5 GHz
Transmit step size	125 kHz	125 kHz	125 kHz
Receive frequency band	10.95–11.7 GHz	10.95–11.7 GHz	10.95–11.7 GHz
	12.5–12.75 GHz	12.5–12.75 GHz	12.5–12.75 GHz
Receive step size	1000 kHz	1000 kHz	1000 kHz
Size of antenna (m)	1.5×1.5 m	1.8 m	1.5–1.8 m
Transmitter power (W)	300 or 2×300 W	300 or 2×300 W	300 or 2×300 W
EIRP (dBW)	69 dBW (300 W)	70 dBW (300 W)	69–70 dBW (300 W)
	71.5 dBW (2×300 W)	72.5 dBW (2×300 W)	71.5–72.5 dBW
LNB noise temperature (K)	130 K	130 K	130 K
G/T (dBK^{-1})	20.5 dBK^{-1}	22.6 dBK^{-1}	22.6 dBK^{-1}
Elevation adjustment	0–90°	0–90°	0–90°
Azimuth adjustment	$\pm 90°$	$\pm 90°$	$\pm 90°$
Polarisation adjustment	$\pm 90°$	$\pm 90°$	$\pm 90°$
Temperature	from $-25°$ to $+45°$C	from $-25°$ to $+45°$C	from $-25°$ to $+45°$C
Wind	16 m/s	16 m/s	20 m/s
Weight			
Antenna (kg)	56	2×45	56–90
Feed/frame (kg)	48	48	48
Mount (kg)	62	62	62
Electronics 1 (kg)	32	32	32
Electronics 2 (kg)	56	56	56
Total flight case weight (kg)	254	288	254–288
Cost	320 000 DM	340 000 DM	from 390 000 DM

Type of station	Truck mounted	Truck mounted
Equipment name	Midi2.6	Midi1.8
Operating band	Ku-band	Ku-band
Transmit frequency band	14–14.5 GHz	14–14.5 GHz
Transmit step size	125 kHz	125 kHz
Receive frequency band	10.95–11.7 GHz	10.95–11.7 GHz
	12.5–12.75 GHz	12.5–12.75 GHz
Receive step size	1000 kHz	1000 kHz
Size of antenna (m)	2.6 m	1.8 m
Transmitter power (W)	2 × 300 W	2 × 300 W
EIRP (dBW)	75.5 dBW	72.5 dBW
LNB noise temperature (K)	130 K	130 K
G/T (dBK^{-1})	25.1 dBK^{-1}	23.5 dBK^{-1}
Elevation adjustment	0–80°	0–80°
Azimuth adjustment	5–65°	± 90°
Polarisation adjustment	± 90°	± 90°
Temperature	from − 20° to + 50°C	from − 20° to + 50°C
Wind	20 m/s	20 m/s
Cost	1 150 000 DM	950 000 DM

Multipoint Communications

Address: Satellite House, Eastways Industrial Park, Witham, Essex CM8 3YQ, UK
Phone: +44 376 510881
Fax: +44 376 502233

Type of station	Fly away	Fly away
Equipment name	SE9510/9710	SE9510/9710
Operating band	Ku-band	C-band
Transmit frequency band	14–14.5 GHz	5.85–6.4 GHz
Transmit step size	125 kHz	125 kHz
Receive frequency band	10.95–12.75 GHz	3.625–4.2 GHz
Receive step size	125 kHz	125 kHz
Size of antenna (m)	1.5–2.4 m	1.5–2.4 m
Transmitter power (W)	300 − 2 × 300 − 500 − 2 × 500 − 2 × 600 W	300 − 2 × 300 − 500 − 2 × 500 − 2 × 600 W
EIRP (dBW)	69.5–78.6 dBW	62–72.0 dBW
LNB noise temperature (K)	100 K	45 K
G/T (dBK^{-1})	23.2 dBK^{-1}	14.5 dBK^{-1}
Elevation adjustment	0–90°	0–90°
Azimuth adjustment	± 40°	± 40°
Polarisation adjustment	± 100°	± 100°
Temperature	from − 20° to + 50°C	from − 20° to + 50°C
Wind	17 m$s	17 m$s

Weight		
Antenna (kg)	50 kg	50 kg
Feed$frame (kg)	15 kg	15 kg
Mount (kg)	65 kg	65 kg
Electronics 1 (kg)	60 kg	60 kg
Electronics 2 (kg)	50 kg	50 kg
Total flight case weight	240 kg	240 kg

Cost	£ 100 k/200 k	£ 100 k/200 k

Type of station	Van/truck mounted	Van/truck mounted

Equipment name	SE9530/9730	SE9530/9730
Operating band	Ku-band	C-band
Transmit frequency band	14–14.5 GHz	5.85–6.4 GHz
Transmit step size	125 kHz	125 kHz
Receive frequency band	10.95–12.75 GHz	3.625–4.2 GHz
Size of antenna (m)	2.4 m	2.4 m
Transmitter power (W)	$300 - 2 \times 600 - 2 \times 600$ W	$300 - 2 \times 600 - 2 \times 600$ W
EIRP (dBW)	73–76–78.5 dBW	66–69–71.5 dBW
LNB noise temperature (K)	100 K	45 K
G/T (dBK^{-1})	21–24.7 dBK^{-1}	16.7–20.7 dBK^{-1}
Temperature	from $-30°$ to $+50°$C	from $-30°$ to $+50°$C
Wind	75 km/h	75 km/h

Cost	£300 k/400 k	£300 k/400 k

REFERENCES

[ABR92] N. Abramson (1992) Fundamentals of packet multiple access for satellite networks, *IEEE Journal on Selected Areas in Communications*, **10**, No. 2, pp. 309–316.

[ALB93] J. Albuquerque, L. Buchsbaum, C. Meulma, F. Rieger, X. Zhu (1993) VSAT networks in the INTELSAT system, *International Journal of Satellite Communications*, **11**, No. 4, pp. 229–240.

[AMU92] P. C. Amundsen, T. Pettersen (1992) The TSAT system—a novel low cost closed SATCOM network for low data rate communication, *International Conference on Digital Satellite Communications, ICDSC*-92, paper B6.

[BAU91] W. Bauerfeld, H. Westbrock (1991) Multimedia communication with high speed protocols, *Computer Networks and ISDN Systems 23, Elsevier Science Publications*, pp. 143–151.

[BOE93] A. Bottcher, C. Meier-Hedde (1993) Static and dynamic retransmission control for slotted ALOHA in transmission channels with erasure and capture, *International Journal of Satellite Communications*, **11**, No. 5, pp. 271–277.

[BUC92] L. Buchsbaum, N. Kusmiri, W. Karunaratne (1992) Technological developments for the provision of thin-route satellite services using C-band VSAT terminals, *International Conference on Digital Satellite Communications, ICDSC*-92, paper A12.

[BUL93] S. Bull, VSAT market status report (1993) *Proceedings of the 10th European Satellite Communications Conference*, London, pp. 99–107.

[CAC91] E. R. Cacciacami, C. A. Politi (1991) VSATs take LANs airborne, *Networking Management*, **9**, No. 12, pp. 54–58.

[CEL91] N. Celandroni, E. Ferro (1991) The FODA-TDMA satellite access scheme: presentation, study of the system, and results *IEEE Transactions on Communications*, **39**, No. 12, pp. 1823–1831.

[CHA88] D. Chakraborty (1988) VSAT communications networks—an overview, *IEEE Communications Magazine*, **26**, No. 5, pp. 10–24.

[CHI88] D. M. Chitre, J. S. McCoskey (1988) VSAT networks: architecture, protocols and management, *IEEE Communications Magazine*, **26**, No. 7, pp. 28–38.

[CHI89] D. M. Chitre, A. C. Briancon, R. Kohli (1989) Random access with notification—a new multiple access scheme for VSAT networks, *COMSAT Technical Review*, **19**, No. 1, pp. 99–121.

[DOR92] Dornier documentation (1992) *Advanced Business Communications via Satellite (ABCS): system description*.

[ELI93] G. Elineau, K. Unterbusch (1993) New trends in satellite news gathering, *Proceedings of the Third European Conference on Satellite Communications, ECSC*-3, pp. 38–43, Manchester (UK).

[ETS 300 157] European Technical Standard (1992) *Receive Only VSATs used for data distribution operating in the 11/12 GHz frequency bands*, European Telecommunications Standard Institute.

[ETS 300 159] European Technical Standard (1992) *Transmit/receive VSATs used for data communications operating in the 11/12 GHz frequency bands*, European Telecommunications Standard Institute.

[ETS 300 160] European Technical Standard (1992) *Control and monitoring at a VSAT*, European Telecommunications Standard Institute.

[ETS 300 161] European Technical Standard (1992) *Centralised control and monitoring for VSAT networks*, European Telecommunications Standard Institute.

[ETS 300 193] European Technical Standard (1993) *General requirements of the connection of VSAT systems to terrestrial networks*, European Telecommunications Standard Institute.

[ETS 300 194] European Technical Standard (1993) *The interconnection of VSAT systems to packet switched public data networks*, European Telecommunications Standard Institute.

[EUT92] (1992) *EUTELSAT 2 Handbook*, Eutelsat, Paris.

[EVA90] B. G. Evans (ed.) (1990) *Satellite Communications Systems*, IEE London, 2nd edition.

✓ [EVE92] J. Everett (ed.) (1992) *VSATs: Very Small Aperture Terminals*, IEE Telecommunications series 28, Peter Pelegrinus.

[FAI93] G. Fairhurst, A. Z. M. Salleh, P. S. Wan, N. Samaraweera (1993) A WAN interworking unit for satellite links, *Proceedings of the Third European Conference on Satellite Communications (ECSC-3)*, Manchester, pp. 187–191.

[FER90] D. Ferrari (1990) Client requirements for real time communications services, *IEEE Communications Magazine*, pp. 65–72.

[FUJ86] A. Fujii, Y. Teshigawara, S. Tejima, Y. Matsumoto (1986) AA/TDMA—Adaptive satellite access method for mini-earth station networks, *GLOBECOM 86*, paper 42.4, Houston.

[FUJ93] A. Fujii (1993) VSATs in Japan, *International Journal of Satellite Communications*, **11**, No. 4, pp. 217–222.

[GAR93] J. Garland, S. Irani (1993) A Canadian demonstration system in support of enhanced personal and business services, *Proceedings of the Third European Conference on Satellite Communications (ECSC-3)*, Manchester, pp. 133–145.

[HA86] T. T. Ha (1986) *Digital satellite communications*, Macmillan.

[HAR92] R. A. Harris (1992) Transmission and access techniques for LAN interconnection, *Proceedings of the 9th International Conference on Digital Satellite Communications (ICDSC92)*, Copenhagen, pp. 179–186.

[ICO93] ICOM Strategies (1993) *VSAT buyer's kit*.

[ITO92] Y. Ito (1992) Standardization of VSAT communication systems, *Proceedings of the 9th International Conference on Digital Satellite Communications (ICDSC92)*, Copenhagen, pp. 109–116.

[ITO93] Y. Ito and P. Amadesi (1993) Standardization of VSAT communication systems in the CCIR, Special Issue on VSAT standardization, *International Journal of Satellite Communications*, **11**, No. 4, pp. 173–179.

[ITU88] (1988) *Handbook on Satellite Communications*, International Telecommunication Union, Geneva.

[ITU90] *Radiocommunications Regulations* (1990) International Telecommunication Union, Geneva.

[JEN80] Y. C. Jenq (1980) On the stability of slotted ALOHA systems, *IEEE Transactions on Communications*, COM-28, pp. 1936–1939.

[JOH 92] J. T. Johnson (1992) Users rate VSAT networks, *Data Communications Magazine*.

[JON88] L. Jones (1988) VSAT technology for today and for the future—Part V: planning and implementing the network, *Communications News*, **25**, No. 2, pp. 44–47.

[KLE75] L. Kleinrock, S. S. Lam (1975) Packet switching in a multiaccess broadcast channel: performance evaluation, *IEEE Transactions on Communications*, COM—28, pp. 410–423.

[KOB90] K. Kobayashi, J. Namiki, A. Kanemasa, R. Saga, K. Watanabe (1990) A satellite LAN interconnecting system, paper 17.4, *IEEE International Telecommunications Symposium ITS90*, Rio de Janeiro.

[LEL91] W. E. Leland, D. V. Wilson (1991) High time resolution measurement and analysis of LAN traffic: implications for LAN interconnection, *Proceedings of the IEEE INFOCOM'91*.

[LEL93] W. E. Leland, M. S. Taqqu, W. Willinger, D. V. Wilson (1993) On the self similar nature of Ethernet traffic, *ACM SIGCOMM'93*, p. 183.

[LEN93] T. Le-Ngoc, J. I. Mohammed (1993) Combined free/demand assignment multiple access (CFDMA) protocols for packet satellite communications, *IEEE International Conference on Universal Personal Communications, ICUPC'93*.

[LIA92] L. S. Liang, J. F. Chang (1992) Response time calculation for VSAT networks, *International Journal of Satellite Communications*, **10**, No. 2, pp. 61–70.

[LIN93] N. Linge, E. Ball, J. Ashworth (1993) Achieving network interconnection using satellite services, *Proceedings of the Third European Conference on Satellite Communications (ECSC-3)*, Manchester, pp. 16–169.

[LON91] M. Long (1991) *World Satellite Almanac*, MLE Inc.

[MAR93] G. Maral, M. Bousquet (1993) *Satellite Communications Systems*, Wiley, 2nd edition.

[MAR94] G. Maral (1994) The ways to personal communications via satellite, *International Journal of Satellite Communications*, **12**, No. 1, pp. 3–12.

[MIT93] M. W. Mitchell, R. A. Hedinger (1993) The development of VSAT performance standards in the United States of America, Special Issue on VSAT standardization, *International Journal of Satellite Communications*, **11**, No. 4, pp. 181–194.

[MOH86] F. Mohamadi, D. L. Lyon, P. R. Murell (1986) Effects of solar transit in Ku-band VSAT systems, *Proceedings of the IEEE Global Telecommunications Conference*, GLOBECOM 86, paper 42.5.

[MOH88] F. Mohamadi, D. L. Lyon, P. R. Murrell (1988) Effects of solar transit in Ku-band VSAT systems, *International Journal of Satellite Communications*, **6**, pp. 65–71.

[MOR88] W. L. Morgan, D. Rouffet (1988) *Business Earth Stations*, John Wiley & Sons, Inc.

[MOU92] C. P. Moura, M. C. Rodrigues, W. A. Russo (1992) Comparative analysis of VSAT networks utilizing TDMA and CDMA access methods, *International Conference on Digital Satellite Communications, ICDSC-8*, Copenhagen, pp. 101–108.

[RAY84] D. Raychauduri (1984) ALOHA with multipacket messages and ARQ-type retransmission protocols—Throughput analysis, *IEEE Transactions on Communications*, **COM-32**, No. 2, pp. 148–154.

[RAY87] D. Raychauduri, K. Joseph (1987) Ku-band satellite data networks using very small aperture terminals—Part 1: multi-access protocols, *International Journal of Satellite Communications*, **5**, pp. 195–212.

[RAY88] D. Raychauduri, K. Joseph (1988) Channel access protocols for Ku-band VSAT networks: a comparative study, *IEEE Communications Magazine*, **26**, No. 5, pp. 34–44.

[RES93] P. Resele, O. Koudelka, O. Lendl, U. Hofmann, C. J. Adams, M. Hine (1993)

Satellite transmission and networking Technical report on Work Package 1.1: information gathering, ESTEC contract 10193/92/NL/LC.

[ROB73] L. G. Roberts (1973) Dynamic allocation of satellite capacity through packet reservation, *National Computer Conference*, pp. 711–716.

[ROG86] D. V. Rogers, J. E. Alnutt (1986) System implications of 14/11 GHz path depolarisation—Part 1: predicting the impairments, *International Journal of Satellite Communications*, **4**, pp. 1–11.

[SAL88] S. Salamoff (1988) Real world applications prove benefits, *Communications News*, **25**, No. 1, pp. 38–42.

[SAL93] J. Salomon, S. Bull (1993) VSAT standards, status and applications in Europe, Special Issue on VSAT standardization, *International Journal of Satellite Communications*, **11**, No. 4, pp. 181–194.

[SCH77] M. Schwarz (1977) *Computer Communications Network Design and Analysis*, Prentice Hall.

[SMI72] F. L. Smith (1972) A nomogram for look angles to geostationary satellites, *IEEE Transactions on Aerospace and Electronic Systems*, AES-5, p. 394.

[SON92] N. Sonetaka, R. Saga, E. Okamoto, K. Nakamura (1992) A proposal for cryptosystem for VSAT networks, *AIAA Communications Satellite Systems Conference*, pp. 1315–1319.

[TAN89] A. S. Tanenbaum (1989) *Computer Networks*, 2nd edition, Prentice Hall.

[TUC94] E. Tuck, D. P. Patterson, J. R. Stuart, M. H. Lawrence (1994) The Calling network: a global wireless communication system, *International Journal of Satellite Communications*, **12**, No. 1, pp. 45–61.

[TUG82] D. Tugal, O. Tugal (1982) *Data transmission*, McGraw-Hill.

[WEL93] P. Wells, A. C. Smith, M. W. Holloway, D. P. Robinson (1993) Milpico terminals—Military personal satcom, *Proceedings of the Third European Conference on Satellite Communications (ECSC-3)*, pp. 38–43, 257–262, Manchester (UK).

[YAN92] O. W. W. Yang, X.-S. Yao, K. M. S. Murthy (1992) Modeling and performance analysis of file transfer in a satellite wide area network, *IEEE Journal on Selected Areas in Communications*, **10**, No. 2, pp. 428–436.

[ZEI91] T. Zein, G. Maral (1991) Stabilized Aloha-Reservation protocol in a DAMA satellite network, *Proceedings of the Second European Conference on Satellite Communications*, Liège, Belgium, pp. 121–126, ESA SP-332.

[ZEI91a] T. Zein, G. Maral, B. Jabbari (1991) Guidelines for a preliminary dimensioning of a transaction-oriented VSAT network, *International Journal of Satellite Communications*, **9**, pp. 391–397.

[ZEI95] T. Zein, G. Maral, T. Brefort, M. Tondriaux, *Performance of the CFRA scheme for aggregated traffic*, COST226 symposium, pp. 183–197, Budapest, May 11–12, 1995.

INDEX